*"This book presents a great deal of
very useful information about the
Brazilian ayahuasca churches, as well as
an exhaustive bibliography of the literature
on ayahuasca in multiple languages.
This material should be of great interest
to all who are interested in these unusual
religious movements which are based on a
powerful plant hallucinogen,
with strong healing, therapeutic
and spiritually significant effects.
A wonderful resource and contribution."*

–Ralph Metzner, PhD, author of
Ayahuasca - Sacred Vine of Spirits.

"Bia Labate and colleagues have compiled a comprehensive review of the world anthropological and clinical literature on ayahuasca. For investigators interested in exploring the fascinating field of ayahuasca studies, this book is a valuable source for our current understanding of the effects of this mysterious vine."

–Charles Grob, MD, Director, Division of Child and Adolescent Psychiatry, Harbor-UCLA Medical Center; Professor of Psychiatry and Pediatrics, UCLA School of Medicine and editor of *Hallucinogens: A Reader* and co-editor (with Roger Walsh) of *Higher Wisdom: Eminent Elders Explore the Continuing Impact of Psychedelics.*

Ayahuasca Religions:

A Comprehensive Bibliography and Critical Essays

Beatriz Caiuby Labate
Isabel Santana de Rose
Rafael Guimarães dos Santos

Translated by Matthew Meyer

PUBLISHED BY

*100% of the profits from the sale
of this book will be devoted to
psychedelic psychotherapy research.*

Ayahuasca Religions: A Comprehensive Bibliography and Critical Essays
ISBN 978-0-9798622-1-2 (trade paperback)
Copyright 2008 by
Beatriz Caiuby Labate, Isabel Santana de Rose, Rafael Guimarães dos Santos
First published in Portuguese, 2008 as
Religiões Ayahuasqueiras: um Balanço Bibliográfico

All rights reserved. No part of this work may be reproduced or transmitted in any form by any means electronic or mechanical except as expressly permitted by the 1976 Copyright Act or in writing by the publisher. Requests for such permission shall be addressed to:
Multidisciplinary Association for Psychedelic Studies (MAPS)
309 Cedar Street, #2323, Santa Cruz, CA 95060
Phone: 831-336-4325, Fax: 831-336-3665
E-mail: askmaps@maps.org

Project Editor: Randolph Hencken, MA, BS
English translation by Matthew Meyer
Cover photos: Ayahuasca preparation by Evelyn Ruman
Banisteriopsis caapi vine by Denizar Missawa Camurça
Fruit of *Psychotria viridis* by Débora Carvalho Pereira Gabrich
Book & cover design: Mark Plummer
Text set in Figural Book for the Macintosh

Printed in the United States of America by McNaughton & Gunn, Saline, MI

To Ilana Goldstein,
Edgard Vianna de Santana and
Iara Moderozo

ACKNOWLEDGMENTS

We extend our thanks to all those who participated in this project, working together to make it possible.

To Professor Oscar Calávia Saez, for the foreword and for his support during the book's development.

To Professor Henrique Carneiro, for his help with the grant proposal to the Fundação de Amparo à Pesquisa do Estado de São Paulo (Fapesp) and for his jacket blurb to the Brazilian edition.

To Professor Alberto Groisman, for his jacket blurb in the Brazilian edition.

To Govert Derix, for his participation and co-authorship of the first English version of the chapter "Bibliography of the Ayahuasca Religions."

To our colleagues and professors who graciously read and commented on the articles presented here: Alberto Groisman, Brian Anderson, Christian Frenopoulo, Denizar Missawa Camurça, Ilana Goldstein, Jean Langdon, José Carlos Bouso, Marcelo Mercante, Matthew Meyer, Maurício Fiore, Mauro Almeida, Renato Sztutman, Sandra Goulart, Sérgio Brissac, Stelio Marras, and Paulo Barbosa.

To those who collaborated on the Portuguese translations of the titles published in the Brazilian edition of the book: Ilana Goldstein for the translations of the titles in German and French and Mauro Almeida for revising them;

Suzana Tavares for the translations from Italian; Momo Nakagawa for those from Japanese and Govert Derix for those from Dutch.

To those who sent us references for the "Bibliography of the Ayahuasca Religions" (at the risk of forgetting a few): Arno Adelaars, Bernd Upnmoor, Clara Novaes, Claude Bauchet, Elena Luppichini, Esther Baguley, Frédérick-Bois Mariage, Hajo Sander, Hans Bogers, Henrik Jungaberle, Irene Hadjidakis-van Schagen, Jacques Mabit, José Carlos Bouso, Jens Martignoni, Kenneth W. Tupper, Lucia Gentil, Manuel Villaescusa, Pieter Lemmens, Randall Sexton, Silvio Andreas Rohde, Sébastien Baud, Fabrizio Mancini, Tatsu Hirukawa, Timberê Villas Boas, Tom Roberts, and Walter Menozzi.

To Rick Harlow, for the cover image of the Brazilian edition. To Denizar Missawa Camurça, Evelyn Ruman and Débora Carvalho Periera Gabrich for the cover image of the English edition. To Daniel Mirante, Basienka Deerheart, Gervásio Santo Silva, Helka Lu, Ricardo Prudêncio Moreira, and Yawa Bane Huni Kuin for the images that appear within the book. To Evelyn Ruman, Andréa D'Amato, Débora Carvalho Pereira Gabrich, Denizar Missawa Camurça, Henrique Biondo, João Guedes Filho, Manuel Poppe, to the Departamento de Memória e Documentação do Centro Espírita Beneficente União do Vegetal, and to the site www.mestreirineu.org for providing photographs. To the photographer David Santos Jr. for his collaboration on the selection and editing of photographs.

To our colleagues at the Núcleo de Estudos Interdisciplinares sobre Psicoativos (NEIP, www.neip.com) for the constant exchange of ideas.

To Fapesp and to the Editora Mercado de Letras, for supporting the publication of the Brazilian edition of the book.

To Rick Strassman, for the initial encouragement to undertake this project.

To the Multidisciplinary Association for Psychedelic Studies (www.maps.org), Richard Wolfe and the Cottonwood Research Foundation (www.cottonwoodresearch.org), for their generous support in funding the book's translation. To Randolph Hencken for the project edition of the book, Mark Plummer for the book and cover design and Valerie Mojeiko for assisting during the process. To Rick Doblin for his friendship.

To Matthew Meyer, for his careful translation of this book from Portuguese into English, and for further revision of the text. To Brian Anderson for his revision of the final proof of the book.

To the Trance Research Foundation (www.trance.edu), Julian Babcock, Michael Walker of DREAMWalker Group (www.dreamwalkergroup.com), Rich Doyle, and Brian Anderson, for their donations to assist with the book's publication in English.

And to MAPS, for its generous support in making possible the English edition of this book.

Contents

PREFACE ...15

Foreword (Oscar Calávia Saez) ...19

CHAPTER 1 ...25
Bibliographical Overview of the Ayahuasca Religions

CHAPTER 2 ...53
Comments on the Pharmacological, Psychiatric, and Psychological Literature on the Ayahuasca Religions

Introduction ...53
The Hoasca Project ...56
The UDV adolescent study ...58
Other studies in Brazil and elsewhere ...61
Spanish research: An emerging field ...64
The UDV and biomedical research on ayahuasca ...69
Health sciences, social sciences, and users: Crossed perspectives ...70
Future prospects ...73
Bibliographical references ...78

CHAPTER 3 ...93
Bibliography of the Ayahuasca Religions

Danish ...94
Dutch ...94
English ...95
French ...108
German ...109
Italian ...113
Japanese ...114
Norwegian ...115
Portuguese ...115
Spanish ...146

ABOUT THE AUTHORS ...151

ABOUT THE PUBLISHER ...154

The tree of life, by Ricardo Prudêncio Moreira – ink on paper
Credits: Ricardo Prudêncio Moreira

PREFACE

The last two decades have seen a broad expansion of the ayahuasca religions, by which we mean the Brazilian Amazonian religious movements that make the use of ayahuasca a central aspect of their ritual. These include Santo Daime (with its Alto Santo and Cefluris variants), União do Vegetal, and Barquinha. The national and international expansion of Cefluris and União do Vegetal (UDV) sparked the development of a multitude of small ayahuasca-using groups in Brazil's large cities, and it has also witnessed, especially since the millennium, a true boom in studies of these religions. This project grew out of the need for an ordering of the profusion of titles related to this subject that are now appearing. We have hoped, in this way, to establish a dialogue between researchers, between lines of research related to the theme of the ayahuasca religions and to offer a vision of a set that we imagine constitutes a field of research. This publication offers a map of the global production of literature on this theme. We hope it will also serve as a reference to researchers in the area and prove useful to others interested in the subject.

For one year, four researchers located in different cities (Beatriz Caiuby Labate in São Paulo; Rafael Guimarães dos Santos in Barcelona; Isabel Santana de Rose in Florianópolis, Brazil; and Govert Derix, in Amsterdam) worked in a virtual research group to compile a list of bibliographical references on Santo Daime, Barquinha, UDV, and urban neo-ayahuasqueiros, including the specialized academic literature as well as esoteric and experiential writings produced by participants of these churches. An Internet odyssey led us to unfamiliar countries, to virtual relationships and intellectual partnerships, then to further questions and the formation of new research networks, each with its own rituals and methodological routines.

We present the results of this collaboration, entitled "Bibliography of the Ayahuasca Religions," with the intention that it be as complete as possible. Completed in November 2007, the present version includes references in ten languages: German, Danish, Spanish, French, Dutch, English, Italian, Japanese, Norwegian, and Portuguese. Whenever possible, we have given links to additional information about the works from the former blog Alto das Estrelas by Beatriz Caiuby Labate (http://alto-das-estrelas.blogspot.com), now converted to her website (http://bialabate.net). In addition, we invited several of the authors whom we met in the course of our research to publish their texts on the site of our research group, the Center for Interdisciplinary Studies of Psychoactives (Núcleo de Estudos Interdisciplinares sobre Psicoativos – NEIP http://www.neip.info), in order to bring to light inaccessible texts. "The List," as is to be expected, is already out-of-date at birth: even as we write these lines some new publication on the topic is surely being issued.

At the same time we produced two texts commenting on aspects of the bibliography we compiled. In "Bibliographical overview of the ayahuasca religions" we have tried to present a profile of these religious groups, including their history and expansion. At the same time, we have attempted a general assessment of the principal characteristics, tendencies, and perspectives evident in the literature about them. The other text, "Comments on the pharmacological, psychiatric, and psychological literature on the ayahuasca religions," summarizes the most important studies of human subjects in the context of Santo Daime, União do Vegetal and Barquinha, evaluating their results, contributions, and limitations. The article offers, in addition, some preliminary anthropological reflections on biomedical research of ayahuasca.

It is our hope thus to contribute to developing and consolidating a field of studies of the ayahuasca religions.

Good reading!

Yube nawa ainbu miyui (Story of the Enchanted Jibóia [boa constrictor]),
by Yawa Bane Huni Kuin (Kaxinawá) – color pencil on paper

FOREWORD

Ayahuasca is worthy of the rivers of ink, which for some time now it has caused to flow. Invited to add some initial comments to this volume, I would do little other than to confirm the prediction offered by its authors: it is fated to a rapid obsolescence, which will come thanks in good part to the information it offers to all those who are interested in the topic. They are many, and diverse: ethnologists, historians, and anthropologists of religion, health specialists, enthusiasts of the visionary arts, devotees of one of the religions that makes ayahuasca its central sacrament. This volume makes a dispersed aggregate into a set, with its own visible boundaries and tendencies; it offers an outline of the shapes of things, a way to find new paths in the rivers of ink.

But I must say something more about the only aspect of this field in which my professional experience permits me to assert the least competence: ayahuasca in the indigenous context (which is, it hardly need be said, its original context). This subject, with sparse exceptions, is absent from the bibliography presented here. What might be seen as a limitation might better be seen as a necessary limit. A bibliography on the indigenous context of ayahuasca would probably have to include all the literature—hardly sparse—produced about the indigenous Western Amazon and, perhaps, a significant number of studies of other regions in the South American Lowlands, since ayahuasca has extended its field of influence to places as distant as the Guarani villages on the Santa Catarina coast in far southern Brazil. Most important, though, that would be a separate work: this foreword may help to fill this gap, in a very synthetic and partial way, but it also says something about the way this gap contributes to the whole.

What differentiates Amerindian ayahuasca from the universe that has erupted out of it is, in the first place, its indefiniteness. Used far from the Indian villages, ayahuasca means religion, whatever that may mean: be it legal or clandestine, redemptive or dangerous, primitive or fruit of the New Age, it is still religion. In the indigenous context it is much more difficult to say what ayahuasca is. To make it the center of an indigenous religion would be misleading in many cases and in others, completely false. There is no lack of situations in which it is treated in a way similar to those of the ayahuasca religions: the brew is (or was) the central element of key rituals in many groups' lives, a relevant sacrament to a collectivity, or at least the proper field of those specialists to whose hands is entrusted the task of maintaining good order. In many cases it organizes a therapeutic system; however, it would be stretching the term to describe this as religious. Ayahuasca can play a role as a remedy in such cases, working directly on the body of the patient; but it may likewise acquire a didactic value, becoming a diagnostic instrument which, through its use in seeking to identify the source of an illness or to enter into contact with its agents, serves the healer as much as the patient. Or it can become the center of a whole shamanic methodology of training and research: througah drinking ayahuasca the apprentice learns the songs, meets and forms alliances with helper spirits, or makes his body, steeped in the bitter substance of the brew, into a proper body for shamanic work. Quite apart from its therapeutic use, and therefore from its most well-known aspect, it is possible to find ayahuasca used as an intoxicating drink in feasts of song and dance in which sensory enjoyment and mutual seduction between the groups of men and women gathered there trump any other motive. It should not be surprising that alcohol, the drink of the whites, ayahuasca's antagonist, opposed in its attributes and incompatible with it, can at times be too its shadow, heir to a whole wicked lineage of the sacred potion. Ayahuasca has, in effect, a dark side that is sometimes manifested in systems of aggression: ayahuasca can be the vehicle by which the spirits of dead kin communicate their desire for vengeance to the living, and also the mark of this violent retribution. To sit down with a group of recent arrivals to drink ayahuasca and sing could occasionally be a dangerous action, recorded in absences and scars.

In many cases ayahuasca appears as an element, although an important one, within a system of plants (alongside tobacco, peppers, datura or the sap of the *samaúma*) constituting a symbolic system parallel to that of the kitchen, in which ayahuasca occupies the "cooked" pole, that which embodies a civilizing role. Alternatively, ayahuasca may stand for this whole complex, assuming by itself this group of functions, or condensing it into an agent of a civilizing process sui generis: more than an ancient tradition, ayahuasca may be a sign of a reformation in indigenous shamanism. In fact, it would be possible to view ayahuasca as the key to a certain cultural ecumene that extends over much of the Western Amazon: songs, designs, and myths which find, through their differences, a common denominator, as for example in the anaconda, whose constant association with ayahuasca is anything but trivial. If on the local level ayahuasca serves communication with spirits, in the region as a whole it facilitates cultural communication and translation. In another, only apparently opposed, sense, ayahuasca acts in indigenous socio-cosmologies like an "othering key,"[1] something akin to the mirror in the western tradition. If the mirror returns us, inverted, our own image, ayahuasca opens the door to a universe in which the same images are presented with their signs reversed; in which the anaconda, which sees itself as human, also drinks ayahuasca and for its turn may see us—who knows—in the shape of anacondas. It is an inversion not of images but of points of view, which can help us understand others, be they spirits, dead people, or foreigners.

This brings us to another aspect that needs to be emphasized, as obvious as it may appear. Ayahuasca is the center of a whole practice and theory of vision. Let us not understand this in too facile a way. One might easily fall into the temptation to take ayahuasca visions as a shortcut to explaining indigenous cosmology, making it an epiphenomenon of the drink's pharmacological properties. Or, who knows: going further along the same path, one might make ayahuasca into a complementary visual organ that activates other perceptual potentials. Ayahuasca may be much more than this. It could, in fact, represent an equivalent to what perspective meant to European art: a way of articulat-

---

1. This point is a comment on an observation made by Eduardo Viveiros de Castro in a lecture at the Universidade Federal de Santa Catarina, with regard to certain Kaxinawá myths.

ing perceptions and a sense of reality. In the European tradition the notion of perspective served, in the first place, to show the limitations of the senses. The eyes deceive, and this deception can be domesticated and redeployed by the artist, as in illusionist painting—the trompe-l'œil—which suggests depths to flat surfaces or presences where there are only representations. Later, in a kind of philosophical trompe-l'œil, that tradition postulated, beyond illusion, a reality, as it exists, whose common denominator was extension, the set of measurable attributes. Perspective, that is to say naturalistic perspective, was one of the pillars of the epistemic success of nature. Ayahuasca may undergird another perspective, another theory of vision, and of the central vision, not of a dimension complementary to—grandiose though it may be—everyday vision. It is central because it is in this dimension, and not in everyday vision, that ontology is based. In many ways similar to that with which modern science postulates about the world---no longer being based on ordinary sight, we now rely instead on microscopes, telescopes, spectroscopes, and other super-ocular instruments. This is not the place to characterize this vision, but some of its key shifts may be suggested: the contrast between figure and ground instead of the choice between appearance and essence; approximation and distancing in place of concealing and revealing; seeing subordinated to looking [o ver subordinado ao mirar—trans.]. It is hard to know—though the texts presented here must offer good clues—how much this theory of seeing has been transferred from the indigenous world to the ayahuasca religions. A certain doubt always hovers over religious or literary movements that claim an indigenous origin, as though indigenous thought outside the jungle could generate nothing more than hollow evocations. I do not believe that the ayahuasca religions, steeped though they may be in an ultimately Christian metaphysic, are entirely inimical to this logic of the concrete that ayahuasca may potentiate in an extraordinary way.

If the Indians have been perceived as objects of the Christian mission, ayahuasca provides the best example of a counter-mission. Not because the Indians have set themselves to the task of proselytizing, although more than a few shamans leave the village to work their science over the whites, presenting interesting similarities to the proclaimers of the gospel who traveled the inverse path. Much might be said about what this expansion of ayahuasca means to the Indians. But I prefer to think about ayahuasca here as an especially expressive case of indigenous creativity later adopted by other peoples, one whose vitality is sufficient for academics to take it seriously. As the authors emphasize in the preface, the ayahuasca religions are not epiphenomena of a psychedelic agent, as powerful as it may be. They are variations on an indigenous cultural theme that increases in potency the further they extend from it. The articles arrayed in this volume offer an excellent sample of this dialogue, which is not always explicit.

Oscar Calávia Saez
Professor of the Post-graduate Program in Social Anthropology,
Universidade Federal de Santa Catarina
Florianópolis, July 2007

Chapter 1

BIBLIOGRAPHICAL OVERVIEW OF THE AYAHUASCA RELIGIONS[2]

In the last decade we have observed enormous growth in the number of studies of the so-called "Brazilian ayahuasca religions," an anthropological category which appeared for the first time, it would seem, in the book *O uso ritual da ayahuasca* (Labate and Araújo 2002 and 2004), and which refers to religious movements of Brazilian origin that have as one of their foundations the ritualized use of ayahuasca: they include Santo Daime, União do Vegetal (UDV) and Barquinha, in their varied denominations.[3]

Faced with this expansion and with the difficulties of locating many of these materials, we compiled a list of bibliographical references on the topic, which aims to be as complete as possible, and to provide a useful guide to researchers in this area.[4] In this article, we offer an overview of these religious groups, and an evaluation of the current state of the global literature about them.

In Brazil, as recently as thirty years ago ayahuasca was very little known, and carried a mysterious aura linked to the "exotic" cults of the faraway Amazon jungle. Beginning in the 1970s these religions were "discovered" by hippies,

---

2. The bibliography for this text can be found in chapter three of this volume.

3. This term has been used, above all, by academic researchers to speak of these groups as a whole, and is not necessarily used by the groups themselves, which often see themselves as unique and independent manifestations.

4. See the "Bibliography of the ayahuasca religions" in this volume.

artists, and intellectuals, by people looking for cures and the merely curious. In the 1980s segments of Brazil's urban middle class began to adopt their practices. By the end of the 1980s they had begun to expand overseas. It didn't take long for this bitter-tasting vine to catch the attention of intellectual types, who quickly baptized it a "sacrament" and categorized these folk-derived rituals as "religious." In similar fashion, the national press soon began depicting them in sensationalist terms.

Raimundo Irineu Serra, an Afro-Brazilian native of Maranhão known as Mestre Irineu (1892-1971), founded Santo Daime in Brazil's north region in the early 1930s. In the late 1970s, one of the groups linked to this religious movement expanded throughout the country and, in the 1980s, internationally. "Santo Daime" (or "Daime") encompasses two principal denominational divisions: several groups generically identified as the "Alto Santo line" and a number of others popularly called the "Padrinho Sebastião line."[6]

The centers which self-identify and are known regionally as the "Alto Santo line" are distinct and function autonomously, although they claim a common origin and maintain close relationships. The primary group identified as Alto Santo is the Universal Light Christian Illumination Center [Centro de Iluminação Cristã Luz Universal (CICLU-Alto Santo)], led by the dignitária[7] Madrinha Peregrina Gomes Serra (1937-), Mestre Irineu's widow, and located in Rio Branco (AC). There is only one thesis that addresses a group linked to the set of Alto Santo churches, specifically the Currents of Universal Light Eclectic Center [Centro Eclético de Correntes da Luz Universal (CECLU)] in Porto Velho, Rondônia (Cemin 1998).[8] The principal group of the Padrinho Sebastião line is the Raimundo Irineu Serra Universal Flowing Light Eclectic Center [Centro Eclético de Fluente Luz Universal Raimundo Irineu Serra (Ce-

---

6. The great variety of possible ways of naming the groups makes it difficult to create classifications that are satisfactory to all (whether researchers or natives). We use the category "line" [Port.: linha], which is also used by some of these religious movements (and which might certainly be critiqued) with the goal of discussing a number of these religious centers together, since without some type of generalization it would be necessary at the extreme to discuss each group separately, as they all have their particularities (for a discussion of the concept of "lines," see Goulart 2004).

7. In the Santo Daime sects we can identify several titles corresponding to ritualistic functions, such as comandante, presidente, fiscal, puxadeira or puxadora, etc. The title of Dignitária belongs only to Peregrina Gomes Serra, also called Madrinha Peregrina, and occurs only in this group—which, in the symbolic economy of this field, represents itself and is identified by many as the "unique" or the "true" Alto Santo, the "purest," "original," or the "root."

8. Sandra Goulart (2004) analyzes several groups that are recognized as being from Alto Santo, giving a good panorama of the set, although she does not center on any one group. Matthew Meyer (2006) is developing the first ethnography of CICLU-Alto Santo of which we have notice.

fluris)],[9] headquartered at Céu do Mapiá (Amazonas state), and whose highest authority is Padrinho Alfredo Gregório de Melo (1950-), son of Sebastião Mota de Melo (1920-1990), or Padrinho Sebastião, and of Rita Gregório de Melo (1925-), also known as Madrinha Rita.

While Alto Santo remained practically restricted to the state of Acre, isolated in the North region, where it numbers today some 800 members (Goulart 2004; the figure is from 2002), Cefluris expanded nationally and internationally. There are currently about forty-two churches in Brazil linked to this group (www.santodaime.org). Overseas, according to preliminary research by Beatriz Caiuby Labate, in 2005 Cefluris had branches in the United States, Canada, and Mexico; one group in Central America, three in South America, twelve in Europe (the most important groups being those in Holland and Spain); and two each in Africa and Asia, for a total of at least twenty-three countries with at least one church. It is important to emphasize, however, that the only Daime churches outside of Brazil that are formally recognized by state authorities are those in Spain and in Holland. According to official figures, Cefluris currently has about four thousand members, Brazilian and foreign.[10]

Cefluris's expansion is also reflected in research on the ayahuasca religions, the great majority of which focuses on this organization (about thirty of the forty-two Brazilian master's theses and doctoral dissertations related to the topic).

With the exception of short references in books such as those by Abguar Bastos (1979) and Nunes Pereira (1979), which are not principally about

---

9. At the Tenth Meeting of the Daimista Churches, held at Céu do Mapiá (Pauini, AM) in 1998, Cefluris was divided into two distinct organizations: one responsible for the spiritual and the religious dimension, and the other for social, environmental, and administrative affairs. The first came to be called the Church of Eclectic Worship of the Flowing Universal Light—Sebastião Mota de Melo, Patron [Igreja do Culto Ecléctico da Fluente Luz Universal – Patrono Sebastião Mota Melo], while the second was named the Environmental Development Institute-Cefluris [Instituto de Desenvolvimento Ambiental Ida-Cefluris]. This Non Governmental Organization's principal objective is the creation and development of sustainable agro-extractivist communities in the areas of the Purus, Juruá, and Pauini rivers and of the Mapiá stream (www.santodaime.org). The new religious name, however, does not appear to be very popular within the group which, as we observed, still uses the old name in the day-to-day. Interestingly, academic writers have also not adopted the new designation, but refer generically to "Cefluris"—perhaps an indication that this term is becoming a kind of anthropological category used to refer to the vast network of Daimista churches and communities that accept the leadership of Sebastião Mota de Melo, in opposition to the set of churches commonly known as Alto Santo.

10. It should be kept in mind that many of these centers are little more than small meetings of groups of friends and relatives; some function overseas in a clandestine or semi-clandestine way; many form and then quickly break into new divisions; many people attend rituals without being formally part of Cefluris, and there is a lot of turnover in participants. All of this makes it difficult to provide a precise count of the number of people regularly involved with this religious organization.

ayahuasca but mention some ayahuasqueiro cults in the Amazon region, the pioneering works on Santo Daime date to the 1980s. The first academic publication is the master's thesis in anthropology by Clodomir Monteiro da Silva, "O palácio de Juramidam – Santo Daime: um ritual de transcendência e despoluição" [The Palace of Juramidam—Santo Daime: A ritual of transcendence and purification,] defended at the Universidade Federal de Pernambuco in 1983. This author argues that Santo Daime's emergence served the needs of the region's urbanizing population and he attempts to show how the appearance of the first Daime groups is linked to regional and inter-regional migratory movements and to the nation's expansionist agenda. The same decade saw the publication of the first books about Santo Daime. In 1983, "História do povo Juramidam: introdução à cultura do Santo Daime" [History of the Juramidam people: Introduction to the culture of Santo Daime] by the historian Vera Fróes Fernandes won a prize sponsored by the Superintendência da Zona Franca de Manaus (SUFRAMA) [the agency that manages development in and around Manaus's Free Trade Zone], and was published as a book in 1986, followed by a second, enlarged edition in 1989. In 1984 there appeared *O livro das mirações: viagem ao Santo Daime* [translated as *"Forest of Visions"*] by the writer, ex-political militant and spiritual leader Alex Polari de Alverga, which became a touchstone for the large urban audience interested in ayahuasca. There followed, in 1989, "Santos e Xamãs" [Saints and Shamans] an MA thesis in anthropology by Fernando de La Roque Couto, who is also the leader of the Daime church Céu do Planalto in Brasília (DF), one of the first churches to be created outside Amazonia. This author analyzes the use of ayahuasca as a type of folk medicine, and more specifically, the "socio-therapeutic use" of the drink within Santo Daime, arguing that this religious movement is characterized by what he calls "collective shamanism."

Other works destined to become important references in the field were produced in the early 1990s. Such is the case of the seminal book by Anglo-Brazilian anthropologist Edward MacRae, *Guiado pela lua – xamanismo e uso ritual da ayahuasca no culto do Santo Daime* (1992) [*Guided by the moon—Shamanism and the ritual use of ayahuasca in the Santo Daime cult*],[11] which emphasizes the importance of a sacred "setting" in the production and control of the social

---

11. This book has been translated to English and published online in 2005 on the website of the Núcleo de Estudos Interdisciplinares sobre Psicoativos (NEIP) (http://www.neip.info/downloads/t_edw2.pdf).

and individual effects of the ingestion of ayahuasca. This book was the first to open a dialogue between religious and shamanic studies and discussions of harm reduction approaches to the consumption of psychoactive substances, and remains a central point of reference in the field. Also from this period is the MA thesis by the Brazilian anthropologist Alberto Groisman, "'Eu venho da floresta': ecletismo e práxis xamânica daimista no 'Céu do Mapiá'" ["I come from the forest": Eclecticism and Daimist Shamanic Practice in "Céu do Mapiá"], defended in 1991 and published as a book in 1999. Groisman's book is also an important reference in the way that it relates the principal categories of Daime cosmology to other manifestations of Brazilian religiosity and to Amazonian shamanism. Also from this period come several other MA theses: Walter Dias's (1992) study "O império de Juramidam nas batalhas do Astral – uma cartografia do imaginário no culto ao Santo Daime" [The Empire of Juramidam in the Battles of the Astral: A cartography of the Santo Daime cult imaginary], which aims to construct a cartography of the Santo Daime imaginary, focusing on the process of expansion to Brazil's urban areas; "A 'Lua Branca' de Seu Tupinambá e de Mestre Irineu: estudo de caso de um terreiro de Umbanda" [The "White Moon" of Seu Tupinambá and of Mestre Irineu: Case study of an Umbanda terreiro] by Maria Beatriz Lisboa Guimarães (1992), the first study to examine the insertion of elements from Umbanda ritual and cosmology into Cefluris; Argentinian doctor Maria Cristina Pelaéz's MA thesis in anthropology, "No mundo se cura tudo: interpretações sobre a "cura espiritual" no Santo Daime" [All is Healed in the World: Interpretations of the "spiritual cure" in Santo Daime] (1994), which discusses Daime conceptions of cure, health, and illness; and the anthropology MA thesis by Sandra Lucia Goulart, "Raízes Culturais do Santo Daime" [Cultural Roots of Santo Daime] (1996), which presents a foundational cultural and historical mapping of this religion, especially with respect to its origins in Amazonian and popular Catholic culture. Finally, the 1990s also saw the publication of two more books by Alex Polari: *O guia da floresta* [*The Guide from the Forest*] (1992) which, like his first book, performs translations and mediations that draw together the universe of Santo Daime and the symbolic world of the new urban members; and *O Evangelho Segundo Sebastião Mota* [*The Gospel According to Sebastião Mota*] (1998), which gives an account of Polari's experiences with Padrinho Sebastião and includes transcripts of his teachings.

The Centro Espírita Beneficente União do Vegetal [Union of the Vegetal Beneficent Spiritist Center (CEBUDV)], also called the União do Vegetal, or simply UDV (www.udv.org.br) was also created in the northern part of Brazil, in Porto Velho, Rondônia, in 1961. It began with José Gabriel da Costa (1922-1971), a native of Bahia know as Mestre Gabriel. According to Edson Lodi, UDV coordinator of institutional relations, this religious movement currently has about fifteen thousand members, or more than triple the membership of Cefluris (personal communication, August 2007). Outside of Brazil, the UDV has "nuclei" or centers in Madrid, and in several places within the United States of America (New Mexico, California, Colorado, Washington, Texas, and Florida), with a total of about 140 members in the US. There are unofficial reports of incipient groups forming in Italy, Portugal, England, and Germany.

The first academic work on the UDV was published in the 1980s, beginning with a pioneering article by Anthony Henman that was published in the Mexican journal *América Indígena* in 1986. That issue of the journal collects articles first presented at the symposium "Chamanismo y uso de plantas del género Banisteriopsis en la hoya amazónica," organized by Luis Eduardo Luna during the XLV Congreso Internacional de Americanistas, held in Bogotá in July 1985. Henman situates the UDV in the context of Brazil's military regime, arguing that the constant threat of repression experienced by the UDV during this time influenced its ideological content. In the 1990s the first theses were published on this group, which even today has been little studied. The religious studies thesis of Afrânio Patrocínio de Andrade (1995) analyzes the belief system of the UDV, pointing to possible motives for its adoption by the urban middle class. The work of Sérgio Brissac (1999) was the first anthropology thesis on the UDV. Brissac's ethnography attempts to capture the central organizing features of the UDV through the study of a nucleus in Campinas (São Paulo state) called Alto das Cordilheiras. He focuses on the group's organizational model, as well as its historical narrative and members' experience of its symbolism. Another reference here is the paper written by Luis Eduardo Luna (1995) in support of his candidacy for a position at the Universidade Federal de Santa Catarina, which compares the UDV and the Barquinha, and for the first time establishes links between the UDV and Peruvian vegetalismo ayahuasqueiro.

Among the Brazilian ayahuasqueiro groups (or hoasqueiro, as the UDV would have it), the UDV is the most self-consciously organized; they are also secretive—or "discreet," as they would prefer to be seen—and are reticent to open themselves up to journalists and researchers, above all those in the social sciences. The UDV is also stricter in its selection and evaluation of new members, who must decide whether to join the group after an initial period of attending rituals. In practice, this situation has resulted in very little academic work on the UDV, which as far as we have been able to discover, has been the subject of just six MA theses in Portuguese, with four in the social sciences (Andrade 1995; Brissac 1999; Luz 1999; Carvalho 2005)[12] and two in the health sciences (Labigalini 1998; Doering-Silveira 2003). Lastly, there is a chapter on the UDV in the PhD dissertation (in social sciences) written by Sandra Goulart (2004), which presents a comparative study of Santo Daime, the UDV, and the Barquinha. This is the only study we know which takes a comparative view of these three groups, and is an important reference in the field of study.

In 2004 the UDV signaled a new willingness to open itself up to researchers in the social sciences when it created the Science Commission, a special agency whose purpose is to work with academics interested in investigating the group. The Commission appears to be the direct result of the increasing academic interest the UDV has stimulated, which seems poised to continue growing. According to our observations, studies have been permitted only so long as their focus is of interest to the group, and their text is submitted for review before publication. Independently of the developments spurred by the Commission's creation, our research indicates that the difficulties confronting studies of the UDV seem to be receding lately. We are aware of at least two ongoing studies in Brazil, one an MA project in the social sciences (Ricciardi 2007), and one a PhD in social anthropology (Melo 2006), in addition to some projects that are still in the UDV Scientific Commission's evaluation process, and therefore were not part of this review.[13] Together, these projects will generate a significant increase in the production of knowledge about this

---

12. There is also an undergraduate thesis by Brian Anderson (2007), the only one to our knowledge published in English on the UDV.

13. According to Luiz Fernando Milanez (personal communication, November 2007), coordinator of the UDV's Scientific Commission, from April 2004, when the commission was created, to November 2007, it evaluated approximately forty requests for research permission presented by scientists from various disciplines

Standards of Mestre Irineu and of the couple Padrinho Sebastião Mota de Melo and Madrinha Rita Gregório Melo. Credits: Art and photos HelkaLu

less-studied religious group.

While there is less known about the UDV from a human sciences perspective than, say, Cefluris, the UDV is nevertheless the leading subject among the ayahuasca religions in biomedical research.[14] The group is the one with greatest investment in creating medical-scientific knowledge about ayahuasca in order to legitimate its ritual use, and has even created an administrative unit to promote this goal, the Medical-Scientific Department (DEMEC). Most of the biomedical and pharmacological research projects conducted in the context of the ayahuasca religions was done within the UDV. The first of these was the Hoasca Project: The Human Pharmacology of Hoasca, conducted in Manaus, in Brazil's Amazonas state, in the 1990s. The study involved nine groups working together—including universities and other institutions located in Brazil, the United States, and Finland—and more than thirty researchers. Fifteen members of the UDV with more than ten years' experience participated as volunteers in the project.[15] Another important pharmacological study, by Evelyn Doering Xavier da Silveira (2003), was also conducted within the UDV, and had as its focus a neuropsychological evaluation of forty adolescents who belonged to the group. The author of the study, a Brazilian psychiatrist, collaborated with foreign scholars on the project, resulting in a series of papers published in a special, international edition of the *Journal of Psychoactive Drugs* dedicated to ayahuasca.[16]

Like the Alto Santo tradition, the Barquinha tradition has little tendency toward expansion, and is the smallest ayahuasca religion in numbers of members, with about five hudred (Goulart 2004; the figure is from 2002). It was created in the rural area around Rio Branco in 1945, founded by Daniel Pereira de Mattos (1888-1958), also known as Frei [Friar] Daniel: The church later subdivided into at least six groups, which remain practically limited to the city of Rio Branco.

---

14. See "Comments on the pharmacological, psychiatric, and psychological literature on the ayahuasca religions" in this volume.

15. See Callaway et al. 1994, 1996; Grob et al. 1996; McKenna et al. 1998; Callaway et al. 1999; Lima 1996-1997; Lima et al. 1998, 2002; Andrade et al. 2004; Brito 2004; Grob et al. 2004. For more information on the Hoasca Project, see "Comments on the pharmacological, psychiatric, and psychological literature on the ayahuasca religions" in this volume.

16. See Doering-Silveira, 2003; Doering-Silveira et al., 2004; Da Silveira et al., 2005; Dobkin de Rios et al., 2005; Doering-Silveira et al., 2005. For more information on these projects and their contexts of production see "Comments on the pharmacological, psychiatric, and psychological literature on the ayahuasca religions" in this volume.

The literature on the Barquinha is also rather limited. The first academic reference is the book by Wladimyr Sena Araújo, *Navegando sobre as ondas do Daime: história, cosmologia e ritual da Barquinha* [Navigating on the Waves of the Daime: Barquinha history, cosmology, and ritual] (1999), which was a published version of Araújo's master's thesis in anthropology (Araújo 1997), and is the only anthropological study of these groups published thus far in Brazil. Araújo's book proposes a hermeneutic reading of the Barquinha's sacred and ritual spaces. There are just two other MA theses in Portuguese on the Barquinha (Oliveira 2002 and Paskoali 2002). There is also a chapter about the Barquinha in the PhD dissertation by Sandra Goulart (2004) already mentioned. Strangely, while the Barquinha group is so little known and relatively small, there are already three academic works in other languages about it: one by the German scholar Carsten Balzer (1998) later published as a book (2003), and two in English; one of them an MA thesis in anthropology (Frenopoulo 2005) and the other, a PhD thesis in human sciences (Mercante 2006). The situation is paradoxical. Despite the small number of studies of the Barquinha, in proportion to its numerical and geographical expansion, it may be the most studied of the ayahuasca religions. Without doubt, there is an element of speculation in this scenario, but we suggest that one of the motives for this interest in the Barquinha is the sheer exuberance of its symbolism and the considerable presence in its rites of Afro-Brazilian elements (especially in comparison with Santo Daime and the UDV), a theme whose study is well established.

The transnational expansion of the UDV and Santo Daime, despite its importance, has been one of the least explored areas in studies of the ayahuasca religions. The first ethnography on the topic is the doctoral work of Alberto Groisman at the University of London, "Santo Daime in the Netherlands: An anthropological study of a New World religion in a European setting" (2000), on Dutch Daime groups. This study has the merit of being the first on the theme, but does not address the political, ethical, and legal difficulties that the expansion of these religions has raised, topics that should get more attention in the coming decades. The legal question is the central issue in Groisman's post-doctoral work (Groisman 2006), in which he discusses legislation relevant to ayahuasca and the legal and political wrangling involved with the regulation of Santo Daime in the United States.[17] There is also an article by Carsten Balzer about Santo Daime in Germany, which takes into account only the initial period of the group's adaptation to that country (1999, 2004, and 2005). The Spanish psychologist Manuel Villaescusa (2002) discusses Santo Daime rituals in England, but there is no consideration in his work of the cultural aspects of the process of international migration of this religious movement, or its effect on participants' subjectivity.

The emergence of "dissident" groups is another important development in the expansion of the ayahuasca religions, one through which new urban forms of ayahuasca consumption are created—mixed with the New Age movement, holistic therapy, various orientalisms, art forms (such as painting, theater, and music), and even intervention programs for the homeless—in what Beatriz Caiuby Labate (2004) has called an "urban ayahuasca network." These "urban neo-ayahuasqueiros" maintain an ambiguous relationship with the religious matrixes from which they are derived. On one hand, they claim a historical and symbolic connection with such groups which gives them a certain legitimacy; on the other hand, these groups tend to reject ayahuasqueiro religious models that they see as "traditional." As they construct new rituals and sets of doctrinal references, they strive to avoid falling into something that could be characterized as "profane drug use." In the next few decades ethnographies of these groups will likely appear. An estimate of the total number of persons currently involved in the various organizations that use ayahuasca would have to include these new independent groups, which have not stopped growing and subdividing. A survey published by Labate (2004) counted twenty-one groups in the city of São Paulo; the current number may be more than double that estimate.

From the late 1990s onward, a progressive expansion of studies about these religions of Amazonian origin could be noted. According to our research, there are currently fifty-two Brazilian academic publications about Santo Daime, the Barquinha, the UDV, and their splinter groups. The great majority of these (35) consists of master's theses; there are just seven PhD dissertations, and one post-doctoral project (Groisman 2007). In addition, there are

---

17. It is very likely that more research will be published in the future on the legal status of the ayahuasca religions. Besides the work of Groisman (2006-2007) already mentioned, incipient examples may be found in: Van der Plas (2002), Labate (2005), Meyer (2005, 2006), Godoy (2006), Groisman e Dobkin de Rios (2007), Koliopoulos (2006) e Kürber et al. (2007).

now nine projects underway (three MA projects, five PhD, and one post-doctoral), not counting undergraduate senior thesis projects, papers presented at conferences, reports and books written by members of the religions, in addition to a series of documentary films, amateur and professional, as shown in the following table. It is also important to note the interdisciplinarity of the field: the literature represents at least eleven different areas, including anthropology, social sciences, history, religious studies, communication, psychology, mental health and psychiatry, education, music, ecology, and hospitality studies. The largest group of studies is in the area of anthropology, where there are eight MA theses, one PhD, and two projects in progress. Also of note is the rapid proliferation of internet sites dedicated to the ayahuasca religions, the better part of them put together by the groups themselves. This is an aspect that was nonexistent ten years ago.

Overview of the Research in Brazil 1

Master's Theses…	35
Doctoral Theses…	07
On-going Research Studies…	09

The 1990s also saw the first important events held in Brazil with the ayahuasca religions as their focus. The Encontro de Estudos sobre Rituais Religiosos e Sociais e o Uso de Plantas Psicoativas (I ERSUPP) [First Meeting for the Study of Social and Religious Rituals and the Use of Psychoactive Plants], organized by Edward MacRae and held in Salvador, Bahia in October 1995, was probably the first international conference of this type held in Brazil, and brought together Brazilian and foreign researchers, such as Jonathan Ott, Luis Eduardo Luna and Jacques Mabit. That same year, the União do Vegetal sponsored a conference in Rio de Janeiro, the First International Conference of Hoasca Studies. More than eight hundred people, including public officials, researchers, and the lay and religious public, participated in the conference, whose objective was to present the results of the Hoasca Project. Both of these meetings were important events in the formation of a field of research on the ayahuasca religions.

Another important moment for this field of investigation was the I Congresso sobre o Uso Ritual da Ayahuasca (I CURA) [First Congress on the Ritual Use of Ayahuasca]. This event was organized by Beatriz Caiuby Labate and was held at the Universidade Estadual de Campinas (UNICAMP) in Campinas, São Paulo, in November 1997. The congress brought together an international group of researchers, as well as members of the religious movements. Its proceedings were published in the book, *O uso ritual da ayahuasca* (*The Ritual Use of Ayahuasca*), edited by anthropologists Beatriz Caiuby Labate and Wladimyr Sena Araújo (2002, 2004 [expanded and revised second edition]). This volume brought together the work presented at the congress and other new articles, and became an important reference in Brazilian studies of the ayahuasca religions. Although the quality of the articles is uneven, the book offers a wide-ranging overview of studies of ayahuasca and brings together several previously unpublished reports. The book examines the use of the ayahuasca vine by the peoples of the forest (indigenous groups, vegetalistas, rubber tappers, and others) and by the religions formed in Brazil, and also includes the results of some medical and pharmacological studies on the substance. The ayahuasca religions are the focus of a central part of the book, within which Santo Daime (and especially Cefluris) receives the most representation, with less emphasis on the União do Vegetal and the Barquinha. The number of published reviews of this book (in the Portuguese, English, Spanish, Italian, and German languages) gives some measure of its importance.[18] One of them, by the ethnologist Peter Gow, received attention in the international anthropological press when it was published in American Anthropologist.[19]

In addition to academic works, in Brazil there has also been a rapid increase in the number of books written by participants in the ayahuasca religions. Among the books by the *fardados* (members) of Cefluris, besides the pioneering works by Alex Polari (1984, 1992, and 1998) already mentioned, other relevant publications include the books *Bença Padrinho* and *Nosso Senhor Aparecido na Floresta* by Lúcio Mortimer (2000, 2001). Mortimer was a sociology student from Minas Gerais state who was one of the first backpackers to live at Colônia Cinco Mil, the community founded by Padrinho Sebastião in the 1970s. He moved permanently to Amazonia and was an important spiritual leader in Cefluris until his death in 2002. There is also a book, issued

---

18. See http://www.neip.info/l_bia_press.html.

19. See http://www.neip.info/downloads/l_bia1_peter.htm.

in a Brazilian *cordel*-style chapbook edition, which tells a version of the life story of Santo Daime's founder in verse. Called *Raimundo Irineu Serra – Mestre Império Juramidam*, it was written by Ceará state native Zerivan de Oliveira (2006), professor of literature at the Universidade Federal do Ceará and member of the Cefluris-affiliated church Flor do Cajueiro, in Cascavel, Ceará. More recently, Padrinho Alfredo Gregório de Melo, the president of Cefluris, published a pair of books, also chapbook verse narratives: *Viagens ao Juruá* [*Journeys to Juruá*] (2007), a bilingual edition (Portuguese/English) tells of the author's first trips to Juruá; the other book, *Padrinho Sebastião–biografia versejada* [*Versified Biography of Padrinho Sebastião*] (2007), tells the story of the Amazonian rubber tapper and founder of Cefluris, Sebastião Mota de Melo.

There are also books written by leaders and participants of the groups linked to Alto Santo. *Contos da Lua Branca: histórias do Mestre Raimundo Irineu Serra e de sua Obra Espiritual, contadas por seus contemporâneos* [*Tales From the White Moon: Histories of Mestre Raimundo Irineu Serra and of his Spiritual Work, as Told by His Contemporaries*] (2003), by Florestan Maia Neto, leader of a group in the Alto Santo line, has the merit of presenting new historical information and an exegesis of Daime hymns developed through conversations with long-time followers. Saturnino Brito do Nascimento, leader of another group in this line, wrote *No brilho da Lua Branca* [*In the White Moon's Glow*] (2005), a small book which, like the booklet by Oliveira (2006) mentioned above, tells a cordel-style version of the life of Raimundo Irineu Serra in verse. The number of works written by participants in Santo Daime published recently indicates a tendency toward the expansion of literary writing about this religious movement.

In addition, there is also a group of publications either written by ex-members or which present criticisms or accusations of some of these religious groups, among which two stand out: *Santo Daime – fanatismo e lavagem cerebra*, [*Santo Daime—Fanaticism and Brainwashing*] written by an ex-member of Cefluris (Castilla 1995); and *Tragédia na seita do Daime* [*Tragedy in the Daime Cult*] written by the stepfather of a member of the same group. It is possible that such scandalous and sensationalist writing may proliferate in the future, a tendency that has already been observed on various internet sites. Beyond the critical character of some writing, we have observed an increase in a type of publication written by members and founded on empirical research conducted outside the academy, as though by an act of appropriation of scientific discourse, frequently (but not always) with the intention of legitimating certain religious ideologies. Examples that deserve special notice include the virtual books written by Luiz Carlos Teixeira de Freitas (2006a and 2006b), a psychologist, journalist and writer, founder and counselor of the Casa de Oração Sete Estrelas [The Seven Stars House of Prayer], a Daime group located in Cotia, São Paulo state. The author attempts to establish a supposed "original form" of Santo Daime from "Mestre Irineu's time," and distinct from the "perversions" created by Ceflúris above all, but also by CICLU-Alto Santo and by other groups who identify themselves as followers of Raimundo Irineu Serra. Although he draws on fieldwork and offers valuable historical data, including some important questions not addressed by anthropologists in the field (such as older ideas about the nature of "daime" itself), the author distorts the facts and manipulates his analysis in such a way as to criticize other groups and to legitimate his own vision of what constitutes "correct" Daime practice. From a theological perspective, this correct and "original" practice is basically a Christian system, stripped of any other type of influence, such as African, Spiritist, esoteric, or indigenous. Thus, in his writings, and above all in the book *O mensageiro – o replantio daimista da doutrina cristã* [*The Messenger: The Daimist Replanting of the Christian Doctrine*] (2006), Freitas uses an intellectual and sometimes sophisticated rhetoric in the name of "original purity" to conduct a kind of cleansing of Santo Daime. Far from being a peculiarity of this author, this type of demand for "purity" and "authenticity" can be considered a constant in this religious field (Goulart 2004; Labate 2004).

Even as we see growth in the number of books published about Santo Daime that are written by church members, the same cannot be said of the UDV. This absence of firsthand accounts of religious experience by members of the UDV is noteworthy when we keep in mind that this is the largest of Brazilian ayahuasca religions in terms of number of participants. Writing books about personal experiences seems not to fit well with the formal hierarchy and the cultural ethos of the group. There is just one book by a participant in the UDV, *Estrela da minha vida: histórias do sertão caboclo* [*Star of My Life: Stories from the Caboclo Backlands*], by Edson Lodi (2004), a member of the UDV for more than thirty years and an ex-president of the institution. Perhaps this book signals the start of a greater willingness of the UDV to share historical, symbolic, and even experiential aspects of its practice, if the institution's leaders approve it.

There are two books written by Wânia Milanez, participant of a UDV dissident group called Centro Espiritual Beneficente União do Vegetal Luz Paz e Amor [Light Peace and Love Union of the Vegetal Beneficent Spiritualist Center] located in Campinas, São Paulo state, with two branches in that city and another in Mato Grosso state. This group has an ongoing judicial dispute with CEBUDV over the name "União do Vegetal." Milanez, the wife of Joaquim José de Andrade Neto, the group's leader, wrote the first book published by any group self-identifying as followers of Mestre Gabriel, called *O Evangelho da Rosa* [*The Rose Gospel*], which was originally published in 1988 (a third Brazilian edition was published in 2001, and a Spanish translation appeared in 1999). She is also the editor of *Oaska: o misterioso chá da Amazônia – relatos de experiências* [*Oaska: The Mysterious Tea of the Amazon*] (2003), which collects experiential accounts by members of the group, where the Vegetal and the relationships of the disciples with their Mestre appear to work as a kind of cure for various problems and ills. Despite this group's small numbers it has reached a relatively high level of public visibility thanks to its ability to bankroll the publication of various books and the creation of some internet sites, in addition to its involvement in several disputes. These include its service mark battle with CEBUDV, accusations make against the leader of CEBUDV in the United States, and accusations of anti-Semitism stemming from an article published in the magazine Humanus, published by Editora Sama—the publishing arm of the group's leader, led by Milanez.

Members of the Barquinha have written two books: the pioneering *Mestre Antonio Geraldo e o Santo Daime: Centro Espírita Daniel Pereira de Matos Barquinha* [Mestre Antonio Geraldo and Santo Daime. Daniel Pereira de Matos Barquinha Spiritist Center] (1996) by Fátima Almeida, Edimilson Figueiredo, and João de Deus; and *Mestre Daniel: História com a Ayahuasca* (2005), edited by Francisco Lima Margarido and Francisco Hipólito de Araújo Neto in honor of the centennial of Frei Daniel's arrival in Acre, and of sixty years since the Barquinha's founding. This book represents an important effort to gather together local researchers and spiritual leaders, and features texts by Wladimyr

Sena Araújo, Edson Lodi, and Clodomir Monteiro da Silva, among others. A third publication put together by members of the Barquinha is a small booklet issued in 2007 by the members of the Swordfish Prince Charity Works Spiritist Center [Centro Espírita Obras de Caridade Príncipe Espadarte] in honor of the 50th anniversary of the "mission" of the leader of that center, Francisca Campos do Nascimento.

It is also important to emphasize, as pointed out by Beatriz Caiuby Labate (2004), that there is an interpenetration of academic and non-academic work on the ayahuasca religions: a good part of the academic researchers are either members or sympathizers of the groups they study, a point that is not always made explicit; likewise, a good part of the texts by participants in the religious movements are either sociologically inspired or present good material for sociological analysis. This participation of anthropologists in the groups under study, although it is not unique to this field of research, may be considered characteristic of Brazilian research in the area of ayahuasca studies. In fact, the great majority of anthropologists who study the ayahuasca religions end up *fardado* at some point. (In all the ayahuasca religions the "fardamento" implies an official conversion in which the neophyte begins to wear a kind of uniform during the rituals and assumes a series of responsibilities within the group.) This in turn implies a certain level of commitment to the groups studied and to their discourses, at least in theory. The implications of this act of "going native" for the work produced are several, and have been discussed in a few places, such as Labate (2004) and Isabel Santana de Rose (2007), among others.

Another kind of publication is books put together by the religious groups themselves. In this sense, Cefluris has just one official publication, the book *Normas de ritual* [*Ritual norms*] (1997), which is a kind of manual about the structure and significance of rituals, as well as an attempt to standardize its member groups' activities. The União do Vegetal has published two books (CEBUDV 1989 and 1991). *Hoasca: fundamentos e objetivos* [Hoasca: Fundamentals and objectives] is the UDV's first official document and represents an effort to formulate key concepts for a public audience. Significantly, this book appeared following the 1985 classification of *Banisteriopsis caapi* as a controlled substance by the Health Ministry's Medicaments Division (DIMED), and the subsequent questioning of this decision by the Federal Narcotics Council (CONFEN), which resulted in the formation of the first Multidisciplinary Task Force to research the ritual use of ayahuasca.[20] It is important to note that this process was heavily influenced by the action of the UDV which, at the time, sent a petition to CONFEN asking to have the decision revoked. The UDV's second book, *Consolidação das leis do C.E.B. União do Vegetal* (CEBUDV 1991) is aimed at the UDV's practitioners, and includes the group's charter and the rules that govern it. It is currently in its third edition (1993).

In the international landscape of studies of Santo Daime, the Barquinha, and the UDV, Brazil is clearly the country that has produced the greatest number of works on the subject, with a total of fifty-two books, ninety articles published in books, journals, and websites, seventy papers presented at conferences and the already-mentioned fifty-two academic works, as shown in the table below.

Overview of the Research in Brazil 2

Books...	52
Published Papers...	90
PhD and MA theses (*)...	52
Conference Papers...	70

(*) completed and on-going researches

However, the increasing appearance of publications in other countries should be noted. According to our research, thirty-three books have been published outside of Brazil, along with twenty-four articles and thirty-six academic studies about the ayahuasca religions, including undergraduate theses and MA and PhD theses, in addition to six projects in progress. These academic works include four theses (Balzer 1998; Huttner 1999; Baiker 2004; Baguley 2006), two projects in progress (Fiedler 2007; Schmid 2007) and one undergraduate

---

20. In 1985 DIMED listed *Banisteriopsis caapi* as a prohibited substance in Brazilian territory, without due consultation of CONFEN. A short while later, CONFEN formed a working group to investigate the ritual use of ayahuasca in Santo Daime and in the União do Vegetal. At the conclusion of these investigations, in 1987, ayahuasca was removed from the list of controlled substances by DIMED and authorized for ritual use. The legality of the ritual use of ayahuasca was questioned again in 1988 and 1994, but CONFEN maintained its earlier decision to permit the use of the drink in ritual contexts, adding the recommendation that it not be consumed by people with psychiatric problems, by pregnant women, and by minors. In 2004 the National Antidrug Council (CONAD) suspended these last two restrictions and instituted a Multidisciplinary Task Force to study and monitor the religious use of ayahuasca, as well as to research its therapeutic uses. In 2006 this working group released its report, which has still not been published officially. Once again the report sanctions the religious use of ayahuasca and presents a "deontology" for the use of the drink, that is, a set of ethical guidelines which are aimed at regulating its consumption and preventing its misuse.

thesis (Rohde 2001) in German; one thesis (Camargo 2003) and one ongoing project (Lavazza 2007)[21] in Spanish; three theses (Bize 2003; Lamoureux 2001; Novaes 2006) and one undergraduate thesis (Martins-Velloso 2003) in French; one thesis (Menze 2004) and one ongoing project in Dutch (Wuyts 2007); eight theses (Soibelman 1995; Groisman 2000; Quinlan 2001; Villaescusa 2002; Cougar 2005; Frenopoulo 2005; Sulla 2005; Mercante 2006), two undergraduate theses (Mavrick 2000; Anderson 2007) and three ongoing projects (Hernandez 2006; Meyer 2006; Dawson 2007) in English; and three theses (Prascina 1997; Luppichini 2006; Koliopoulos 2006) and one undergraduate thesis (Curuchich n.d.) in Italian. As in Brazil, there are a large number of foreign publications in the discipline of anthropology, but the output encompasses other areas as well, including psychology, psychotherapy, religious studies, social sciences, and health. There is thus a tendency in the study of the ayahuasca religions toward concentration in the area of anthropology, with some emphasis also in the areas of psychology and mental health/psychiatry.

Outside Brazil, the greatest share of publications has been in the United States, where thirteen books, sixty-three articles, and thirteen academic works have been published. This work is marked, however, by a greater emphasis on pharmacology. It is worth recalling the already-mentioned investigations by the Projeto Farmacologia Humana da Hoasca [Hoasca Human Pharmacology Project], and by Evelyn Doering Xavier da Silveira on adolescents in the UDV (2003). This emphasis on pharmacological research in the United States may be linked to the American emphasis on the "hard sciences," which even today are frequently considered more objective and scientific, in invidious comparison with the social sciences. Moreover, this may also be linked to the efforts by the UDV to stimulate research in the biomedical sciences in order to bolster its legitimacy and credibility in the context of disputes over the legalization of its practices in Brazil and in other countries, since the two main investigations cited here were conducted within this group, and with its collaboration.[22]

Another point deserves comment here: our research points to an enormous

---

21. Vitor Hugo Lavazza, as far as we could discover, is the first researcher to write about this subject in Argentina. One of the themes of his work is the expansion of Santo Daime to that country.

hole in English-language publications on Santo Daime, the Barquinha, and the UDV in the area of anthropology. For example, while in France researchers of some renown, or associated with important universities, such as Jean-Pierre Chaumeil (1992) and Patrick Deshayes (2002, 2004, 2006) have written on the subject, in the United States the theme has yet to attract much attention from anthropologists at prestigious universities. In that country, Santo Daime, the Barquinha, and the UDV tend to be studied by researchers affiliated with institutions that define themselves, or are defined in the academic milieu, as "unconventional" and outside the first tier of universities, as in the case of the California Institute of Integral Studies, the Institute of Transpersonal Psychology and the Saybrook Graduate School and Research Center, all of which are in California and which account for five of the eight theses and dissertations in English. By the same token, it was the California Institute of Integral Studies that sponsored, in 2000, the Ayahuasca Conference.[23] As far as we can tell, this was the first international conference on this subject in the United States, and it brought together scholars from diverse fields, including pharmacology, neuroscience, shamanism, and ethnobotany. This conference was aligned with the psychedelic research movement that has gained ground in the less hegemonic spaces of the American academy, and which represents an attempt to approach the so-called "entheogenic drugs"[24] from a holistic and transdisciplinary perspective.

In addition to English-language works by Anglo-American researchers, there are also important texts by Brazilian researchers that have been published in English. The first of these, as far as we can tell, was the MA thesis by Tânia Soibelman (1995), which deals with the contemporary use of ayahuasca among urban groups in Rio de Janeiro. It was followed by an article by Alberto Groisman about healing in Santo Daime, co-authored with the psychiatrist Ari Sell and published in a book edited by Michael Winkelman (1995), in which the authors attempt to show how the Daime concept of healing goes beyond organic events to include the whole human being in her relations with other

---

22. For more information on this research and the contexts of its production, see "Comments on the pharmacological, psychiatric, and psychological literature on the ayahuasca religions" in this volume.

23. See http://www.conferencerecording.com/newevents/auy20sfbg.htm

24. The term entheogen was proposed by Gordon Wasson, Albert Hofmann and Carl Ruck to refer to plants that have been used as sacred tools of ecstasy. The word comes from ancient Greek, with entheos meaning "inspired or possessed by a god" and the suffix geno designating "generation or production" of something. Thus, a possible translation would be "that which produces a divine inspiration or possession" (Labate, Goulart e Carneiro 2005, p. 51).

living beings and with the cosmos. Groisman also wrote three other texts in English: one about Santo Daime in Europe published in the MAPS Newsletter (1998); his PhD dissertation, already mentioned (2000); and an article co-authored with Marlene Dobkin de Rios about the legal case of the UDV in the United States (Groisman and Dobkin de Rios 2007). Edward MacRae has published four texts in English, including articles (1998, 1999), book chapters (2004) and an online version of his book *Guided by the Moon* (2006), published on the site of the psychoactives study group Núcleo de Estudos Interdisiciplinares sobre Psicoativos (NEIP). This translation deserves special emphasis since, as we have noted, this book is a key reference in studies of this subject.

Alex Polari, who as we saw was a pioneer in the area of books written by members of the ayahuasca religions in Brazil, also has a publication in English: *Forest of Visions: Ayahuasca, Amazonian Spirituality, and the Santo Daime Tradition* (1999), which is a translation of his first book on the topic (1984) with an added preface by Jonathan Goldman, head of a Daime church in Oregon state and an important leader of the movement in the United States.[25] There are also two English-language works that were already mentioned: an MA thesis by the Uruguayan anthropologist Christian Frenopoulo, "Charity and Spirits in the Amazonian Navy: the Barquinha Mission of the Brazilian Amazon" (2005) and a PhD thesis by the Brazilian Marcelo Mercante, "Images of Healing: Spontaneous Mental Imagery and Healing Process of the Barquinha, a Brazilian Ayahuasca Religious System" (2006), which, coincidentally, both address the same Barquinha group, the Centro Espírita Obras de Caridade Príncipe Espadarte [Swordfish Prince Charity Work Spiritist Center], or Barquinha da Madrinha Chica, located in Rio Branco, Acre. Unfortunately, these texts remain unpublished. Thus, while we have identified thirteen academic studies in English on the subject of the ayahuasca religions, only nine were carried out by North American researchers. It is worth mentioning too the work of Andrew Dawson, a religious studies professor at Lancaster University, who recently wrote a book on new religious movements in Brazil, *New Era – New Religions: Religious Transformation in Contemporary Brazil* (2007). The book contextualizes the emergence of New Age religiosity in wider processes

---

25. It should be mentioned that there is also a German translation of Alex Polari's second book (Alverga 2007), highlighting the visibility of discussion of the ayahuasca religions in Germany, which we will discuss below.

of social transformation, and includes a chapter about the ayahuasca religions.

In Europe, one of the countries with the greatest production of work on the subject has been Germany: there are seven books, twelve articles and four theses and dissertations. In addition, at the University of Heidelberg there is an emerging circle of researchers, linked to a research center called "Ritual Dynamics." In 2002, researchers from this center began a cultural-psychological project that aims to use qualitative and quantitative methods to study Santo Daime in Europe, as well as participants of other ayahuasca groups led by South American shamans, by their European disciples, or by Europeans working within therapeutic paradigms, in addition to studying individuals who consume ayahuasca at home or independent drinkers who use it in various contexts. The same center organized a conference, held in May 2008, about the expansion of ayahuasca to Europe. The conference focused on cultural, medical, psychological, and legal questions involved with ayahuasca, particularly outside of South America. Leaders of Daime groups in Brazil and Europe helped plan the conference in conjunction with researchers from Heidelberg University.[26] The concentration of research on the ayahuasca religions in Germany is noteworthy when we consider that the legal status of Santo Daime in that country is still not well defined. Another relevant fact is that in Germany there are at least two PhD projects on ayahuasca underway in the area of medicine (Fiedler 2007; Schmid 2007). This situation may indicate a tendency toward pharmacological and biomedical projects in German research.

The countries in which Santo Daime has the largest presence in Europe are Spain and the Netherlands, which have been through difficult legal processes to secure the right to religious liberty. In Spain, there have been six books and seventeen articles written on the subject, but just one MA thesis, by the Brazilian Ilze Andrade Camargo (2003). Her work was based on observation of Brazilian members of the UDV and of patients in the author's clinic who later became members of the UDV, and is, in our view, not an entirely consistent project. It should be noted that a large part of the publications in Spanish are translations of work by Brazilian researchers. Spain was, however, the country that hosted the first European conference on ayahuasca, the V Jornadas Internacionales sobre Enteógenos. Ayahuasca: Tradición, Nuevas Aplicaciones

---

26. See http://alto-das-estrelas.blogspot.com/2007_03_01_archive.html.

y Futuro, held in Barcelona in September 2002 and organized by the Societat d'Etnopsicologia Aplicada i Estudis Cognitius (Sd'EA), directed by the Catalan anthropologist Josep Maria Fericgla. A number of important researchers attended this meeting, such as Jonathan Ott, Jordi Riba, and Jacques Mabit, and there were about two hundred participants.[27]

The Netherlands, in turn, has produced an impressive array of work: a book (Derix 2004); fourteen articles; a PhD dissertation in psychology (Menze 2004) and an MA thesis in anthropology in progress (Wuyts 2007). In addition, in November 2002 the country hosted Psychoactivity III, another important conference on ayahuasca, held in Amsterdam. Attending this conference, in addition to researchers, were about 150 ayahuasqueiros from Europe, America, and Japan. Disciplines represented in the conference included anthropology, psychology, ethnobotany, pharmacology, and included discussions of the legal situations affecting ayahuasca and the ayahuasca religions.[28]

In addition to the countries already mentioned, we were able to locate publications on the topic in other languages, some of them unusual, such as Danish. In Italy, where there has also been legal repression of Daime, there are seven articles, four academic works, and three books (Gioia 1996; Verlangieri 2000; Menozzi 2007), two of which (Gioia 1996 and Verlangieri 2000) were written by church members (with greater or lesser scientific pretension, as occurs in similar cases). In Japan, where several Daime churches have sprouted up in clandestine form, we have noted at least seven publications on Santo Daime, including three books (Nagatake 1995; Akira 2001; Hirukawa 2002). In the following table we provide a summary of the research and publications about the Brazilian ayahuasca religions abroad.

General overview of research abroad

	Books	Papers	Theses
German	7	12	7
Spanish	6	17	2
French	0	7	0
Dutch	1	14	2

---

27. See www.aya-huasca.com.

28. See http://www.maps.org/news-letters/v13n1/13144sav.html and http://istoe.terra.com.br/planetadinamica/site/reportagem.asp?id=219

English	13	63	13
Italian	3	7	4
Japanese	3	4	0
Total	33	124	36

The large number of research projects and books about Santo Daime, the UDV, and the Barquinha (and their splinter groups) in Brazil and abroad is related to the national and international expansion of these religious movements, and indicates the importance that ayahuasca and the phenomenon of the ayahuasca religions are acquiring in the contemporary world. If the first wave of studies of these religions focused on their historical, cultural, and symbolic aspects, their expansion and entrance into the international religious marketplace (including, according to some reports, the commercialization of the drink itself) has shifted the focus more and more toward issues of drugs and drug policy. National and international discussions about the legality of drinking ayahuasca and about the juridical status of the religious groups have helped draw the attention of civil society, the state, and the media to the phenomenon, thus widening the scope of discussion about it and stimulating the production of new academic studies. These studies, in turn, have played a fundamental role in the legitimization of the ayahuasca-using groups.

Cefluris and the UDV have come to be represented more and more, with the blessing of anthropologists and native intellectuals, as genuinely Brazilian religions (Labate and Pacheco, 2005). One center, the Universal Light Christian Illumination Center (CICLU-Alto Santo), the Alto Santo church led by Peregrina Gomes Serra, was even registered as historical and cultural patrimony of Rio Branco and of Acre state by the mayor and state governor, in 2006. Santo Daime and the UDV, each in their own way, have acquired expansionist characteristics, crossing the ocean toward the Old World, jockeying for space with other religions and moving between various symbolic, economic, and cultural frontiers, contributing to the transformation of ayahuasca into a kind of transnational "pan-entheogen."

Studies of the ayahuasca religions have tried to keep pace with the ways the consumption of the substance has diversified, and have themselves grown in number and type of approach. Such studies become relevant to the degree with which they dialogue with classic and contemporary questions in anthropology and in other disciplines, such as: identity construction; hybridization; corporality and embodiment; religious migration; the global expansion of the New Age movement; the relationships between shamanism, medicine, and religion; conflicts between the "traditional" and the "modern" in the construction of authenticity of cultural practices; processes of cultural objectification; links between religions, politics, and culture; conceptions of the "normal" and the "pathological"; construction of social stigma; the relationship of nature and culture; tensions between subject and object, observation and participation in research; the production of religious ecstasy; and so on. This area of research also has great importance for discussions of the consumption of so-called "drugs," for the consideration of prohibitionism, and the utility of harm-reduction approaches. In this way, while some studies may express a certain reductionist tendency by placing much emphasis on ayahuasca as a substance, and highlighting the consumption of Daime or of Vegetal as the only or the central dimension of those religions, consideration of the theme of the ayahuasca religions leads far beyond the question of the substance in itself.

In the same way that it is necessary to relativize the role of the use of ayahuasca in these practices, the category of "ayahuasca religions" should also be put in perspective, as should the notion of a "field of studies of the ayahuasca religions." None of these is a natural given, but all are the historical products of an academic community with a common focus and a shared discourse. The boom in work on Santo Daime, the UDV, and the Barquinha that the following bibliographical inventory indicates suggests a tendency toward the consolidation of this field of research. As we have seen, this field encompasses a wide range of themes and dialogues with numerous disciplines, which may be one of the factors that explain its tremendous growth in the last few years.

From left to right (only those in uniform): D. Lourdes Carioca (standing),
Md. Peregrina Gomes Serra (seated), Mestre Irineu (seated),
Leôncio Gomes (standing),
Raimundo Mendes (standing at far right)
Credits: Photo available at http://www.mestreirineu.org/Mestre05.jpg,
accessed June 3, 2007

Mestre José Gabriel da Costa, founder of the União do Vegetal
Credits: Cícero Alexandre Lopes, Department of Memory and Documentation
of the Centro Espírita Beneficente União do Vegetal

Mestre Daniel Pereira Mattos
Credits: Image courtesy of João Guedes Filho

Chapter 2

Commentary on the Pharmacological, Psychiatric, and Psychological Literature on the Ayahuasca Religions

Introduction

The objective of this text is to provide an overview, both for researchers and for the non-specialist public interested in pharmacological and biomedical studies of psychoactive substances, of the principal studies made in these areas of the so-called ayahuasca religions—Santo Daime (both the Alto Santo and Cefluris denominations), the União do Vegetal (UDV), and the Barquinha[29]

Our starting point was an exhaustive, worldwide survey of work on the topic at different academic levels, including undergraduate, master's, and doctoral and others. The results of this survey are presented in the bibliography that follows this article.[30] Our survey indicates that there exists an enormous difference, in quantitative terms, between studies carried out in the social sciences on the ayahuasca religions and those done in the biomedical sciences. In the area of the humanities we have seen tremendous growth, both domestic and international, in addition to an increase in books by religious practitioners reporting their experiences and reflecting on the theological aspects of these religions. The number of scientific, biomedical studies on the use of ayahuasca—whether in indigenous, mestizo, or urban contexts—is

---

29. For a general introduction to these religions' history, see "Bibliographical overview of the ayahuasca religions," in this volume.

30. See "Bibliography of the ayahuasca religions," in this volume.

31. See "Bibliographical Overview of the Ayahuasca Religions" in this volume.

quite small.[31] While there exist a considerable number of studies on the botanical and chemical aspects of ayahuasca,[32] the first controlled studies of the drink's effects in the areas of biochemistry, physiology, neurology, psychiatry, and psychology date to the 1990s, and are still few in number and limited in various ways. In fact, research on hallucinogens[33] using human subjects was suspended for a period of more than thirty years. Increasing consumption of hallucinogens in the 1960s, especially in the United States, was met with a strong reaction in certain segments of society, which hardened public policies toward these substances and froze scientific research (Grof 2001; Grob 2002).

Here we will describe a little of that history and present a modest overview of the pharmacological, psychiatric, and psychological studies of the ayahuasca religions, highlighting the principal findings and limitations of these investigations. When necessary, we will present some basic concepts used in this area of research. Given that the emerging literature on the subject is quite diverse and heterogeneous in terms of scientific rigor,[34] we have chosen for detailed discussion those studies to which we have access and which seem to us to be the most important.[35] The remainder of the studies are reviewed generically, as a group. In the conclusion, we examine the prospects in this field for future study, calling attention to specific needs for continuing research. An introductory clarification is necessary: we will not enter here into the complex and thorny questions of the nature of medical-scientific research on health in general and on the use of hallucinogens in particular, the validity of the medical and psychiatric questionnaires used in these studies, etc… For a discussion of the modern separation of "nature" and "culture," see Latour (1994); on the attempt of scientific and naturalistic medicine to isolate the "objective" dimen-

---

32. For a review of the historical, botanical, and pharmacological aspects of ayahuasca, of the beta-carbolines, and of dimethyltryptamine (DMT) see Schultes (1986), Ott (1994), and Strassman (2001).

33. The concept of "hallucination" has been widely criticized for evoking a false and illusory perception of reality, often associated with pathology (for a discussion of this concept see Labate et al. [2005]). We have chosen to maintain the term "hallucinogen" despite its limitations since it is still the most widely used term in the biomedical literature.

34. Callaway et al. 1994; Labigalini Jr. and Dunn 1995; Miranda et al. 1995; Callaway et al. 1996; Grob et al. 1996a, 1996b; Lima 1996-1997; Prascina 1997; Labigalini Jr. 1998; Lima et al. 1998; McKenna et al. 1998; Callaway et al. 1999; Cazenave 2000; Barbosa 2001; Quinlan, 2001; Lima et al. 2002; Shanon 2002; Villaescusa 2002; Barbosa e Dalgalarrondo 2003; Camargo 2003; Chaves 2003; Doering-Silveira 2003; Martins-Velloso 2003; Shanon 2003; Villaescusa 2003; Andrade et al. 2004; Brito 2004; Grob et al. 2004; Menze 2004; Shannon 2004; Araújo 2005a, 2005b; Barbosa et al. 2005; Carvalho 2005; Costa et al. 2005; Cougar 2005; Da Silveira et al. 2005; Dobkin de Rios et al. 2005; Doering-Silveira et al. 2005a, 2005b; Morais 2005; Sulla 2005; Barbosa 2006; Novaes 2006; Santos 2006; Santos et al. 2007.

35. They are: Grob et al. 1996a, 1996b; Labigalini Jr. 1998; Barbosa 2001; Barbosa and Dalgalarrondo 2003; Do ering-Silveira 2003; Andrade et al. 2004; Brito 2004; Grob et al. 2004; Barbosa et al. 2005; Da Silveira et al. 2005; Dobkin de Rios et al. 2005; Doering-Silveira et al. 2005a, 2005b; Barbosa 2006; Santos 2006; Santos et al. 2007.

sion of "nature" from the "subjective" effects of "culture" in laboratory tests of the production of counter-placebo medications, see Pignarre (1999) and Marras (2002);[36] for an analysis of the construction of the biomedical concept of "chemical dependency," see Ott (1998).

All of the research projects presented here, as sophisticated as their methodologies may be, should be considered in light of the complex relationship between the substance itself, the "set," and the "setting". The substance, because of its chemical structure and the dosage, acts in a certain way upon the human organism. The person, his or her biological makeup, personality, personal history, previous experience with the drug in question, as well as the personal motivations and expectations in relation to the drug, are what characterize the "set". The setting is the context in which consumption of the substance takes place, including: the person administering the drug; the people present during the experience; the physical environment in which the experience happens (decor, music, smells, etc.); and the larger cultural context, especially the values attached to the substance (Grinspoon and Bakalar 1981; Zinberg 1984; Grof 2001). Incidentally, it might be emphasized that the very notions of "substance," "set," and "setting" should be recognized as analytical abstractions, since in practice they cannot be completely separated from one another.

An anthropological analysis of the logic of research projects involving the consumption of ayahuasca in religious contexts would be a great challenge, and is yet to be realized. Our purpose here is to examine the studies in terms of their own parameters; that is, in terms of the rigor they maintain with respect to the assumptions accepted by the scientific community at a particular historical moment. To recognize that these studies utilize a certain kind of scientific discourse that is not identical to the "truth" is not the same thing as saying that they are all problematic and limited. Our task here is to comment on the particulars of each study and its success relative to the goals it sets for itself.

Before discussing these studies, some general comments on their character are in order. There are basically two kinds of projects that have been done in the context of the ayahuasca religions: those conducted with animals and those

---

36. The article by Marras is available on the NEIP site: http://www.neip.info/ downloads/Ratos e homens Final.pdf.

conducted with human subjects. The projects involving humans include both investigations *in loco*, that is, in the context of rituals in the religious centers and in the laboratory, where ayahuasca is administered in a clinical environment. Among the participants in these projects are those with long-term experience with the drink (ten years), those with some experience (2-3 occasions), and others who were taking it for the first time. The studies with long-term drinkers of ayahuasca offer the opportunity to analyze its chronic (long-term) effects. Studies of acute (short-term) effects, in turn, allow the observation of effects of a single dose or of a few doses of ayahuasca within a short period of time, and can be conducted with people who have had a lot of experience with ayahuasca and with those who have had very little.

To our knowledge, the only animal studies conducted in the context of collaborative projects with ayahuasca-using groups were those of Carvalho and associates (Carvalho et al. 1995a; Carvalho et al. 1995b). These were part of the Hoasca Project (see below). However, according to our survey, these have not been published. In this chapter we will discuss only the studies done with human subjects. Our reasoning is that, in general, data from animal studies cannot be directly extrapolated to human beings, particularly because of different metabolisms and the complex and subjective nature of cultural contexts.

The Hoasca Project

It can be broadly stated that two studies stand out in the international scientific community. The first was the Hoasca Human Pharmacology Project (or Hoasca Project), which was born out of the first two health conferences held by the UDV in 1991 (in São Paulo) and in 1993 (in Campinas, São Paulo state). One result of the first conference was a proposal made to the ethnopharmacologist Dennis McKenna recommending that he conduct a broad investigation on the use of ayahuasca in the UDV, which he accepted. This resulted in the Hoasca Human Pharmacology Project.[37] The investigation was a collaboration

---

37. According to Glacus Brito, ex-director of the Medical-Scientific Department (DEMEC) of the UDV and assistant physician in the Department of Immunology at the University of São Paulo School of Medicine, in 1991 the UDV made a proposal to Dennis McKenna to study ayahuasca, stimulated by Jeffrey Bronfman, current Representative Mestre of the group in the USA. Dr. Brito is said to have come to Brazil with the "project under his arm" and seeking support. The "first door" that opened, according to him, was the researcher Oswaldo Luis Saide, of the Department of Psychiatry of the State University of Rio de Janeiro (UERJ); the second was Claudio Miranda of the Department of Psychiatry at the Paulista Medical School (EPM) of the Paulista Federal University (UNIFESP); following this came the support of the Multidisciplinary Center for Chemical, Biological, and Agricultural Research at the Campinas State University (UNICAMP) and of the Department of Medicine at the Federal University of Amazonas (UFAM) (presentation at the 1995 International Conference of Hoasca Studies [Conferência Internacional dos Estudos da Hoasca] in Rio de Janeiro; DVD available at the Laboratório de Imagem e Som at the Anthropology Department of the University of São Paulo [USP]. See http://alto-das-estrelas.blogspot.com/2007/04/disponibilizado-material-sobre.html.)

of nine research institutions, including universities and institutes in Brazil, the United States, and Finland, and involved more than thirty researchers. Although Brazilian institutions and researchers participated in the project, the principal investigators (Charles Grob, Dennis McKenna, and JC Callaway) were foreigners.[38] Data were collected at a UDV nucleus in Manaus, Amazonas state, and comprised botanical, chemical, toxicological, pharmacological, neuroendocrine, clinical, and psychiatric aspects, incorporating results of both human and animal assays. Fifteen members of the UDV, each with at least ten years' experience, were recruited as volunteers.

In 1995 the UDV sponsored the 1st International Hoasca Researchers' Conference in Rio de Janeiro to formally present the results of the Hoasca Project. The principal results were published first in English in prestigious international journals (Callaway et al. 1994, 1996; Grob et al. 1996; McKenna et al. 1998; Callaway et al. 1999) and in internet bulletins (Limat 1996-1997). They were later also published in Portuguese (Grob et al. 1996; Lima 1996-1997; Lima et al. 1998; Lima et al. 2002; Andrade et al. 2004; Grob et al. 2004).[39] According to the researchers involved, the Hoasca Project revealed no organic damage (kidney, liver, circulatory, etc.) in the UDV member subjects; no acute damage to the central nervous system; and an absence of psychiatric disturbances, including those involved in drug dependency, such as withdrawal, tolerance, abusive behavior, and social impairment (Grob et al. 1996a, 1996b; Andrade et al. 2004; Grob et al. 2004). In comparison to the control group[40] (subjects who had never taken ayahuasca) the UDV members appeared more confident, optimistic, relaxed, unworried, uninhibited, alert, energetic, happy, and of calmer temperament (Grob et al. 1996a, 1996b, 2004). Moreover, the data showed improvement in pre-existing psychiatric conditions upon joining the UDV. Of the fifteen volunteer members, eleven had been dependent on alcohol (five of them seriously); another two had a history of depression, and three suffered phobic anxiety. Upon joining the group, all of these disorders disappeared (Grob et al. 1996a, 1996b, 2004).

---

38. For more information see http://www.udv.org.br/portugues/texto/ udv-versao-texto.doc.

39. The articles by Grob et al. (1996), Lima (1996-1997), Lima et al. (1998), McKenna et al. (1998) e Callaway et al. (1999) are available at the UDV site: http://www.udv.org.br/portugues/area_ocre/links/links para 3/5 informacoes.html.

40. The control group is a tool commonly used in biomedical studies and consists of individuals similar, in variables such as age, gender, education level, nationality, and so forth, to the subjects investigated

The Hoasca Project had the merit of being the first systematic biomedical investigation of ayahuasca, and deserves recognition for that fact.

Although the study presented positive psychological and psychiatric findings among long-term consumers of ayahuasca and despite the finding of no physiological harm, these results should be regarded with caution, particularly those presented in the work of Grob et al. (1996a, 1996b, 2004): the sample size was small, the subjects were all male, and the neuropsychological tests were not tailored to achieve the least possible interference of cultural, semantic, and methodological factors (Doering-Silveira 2003). In addition, it could be argued that the exclusive use of long-term members represents a selection bias favoring those who have been able to adapt to the consumption of this psychoactive drink.[41] It should also be emphasized that the study design does not permit the evaluation of the impact of adherence to a religious doctrine on the positive effects reported. Even so, these results indicate possible beneficial effects of ayahuasca, suggesting that future work should attempt to replicate them with more rigorous methods.

The UDV adolescents study

The second principal scientific study of the effects of ayahuasca in humans was conducted with adolescent members of the UDV. According to the *Alto Falante* newsletter published by the UDV, the study was the result of a request made to the group by the Federal Public Ministry.[42] We were unable to obtain further information about this, but several sources claimed that the study was conducted on the initiative of the UDV. The project grew out of the group's questioning of a recommendation by the old Brazilian narcotics agency, CONFEN, that ayahuasca not be given to minors (cf. Labate 2005), and its proposal to several scientists, some time later, to carry out a scientific study of the subject.[43] The results of the study were published in several stages.

---

41. For a more detailed and critical analysis of the results of the Hoasca Project, see McKenna et al. (1998) and McKenna (2004).

42. Here is the newsletter's version: "The research was solicited by the Public Federal Ministry (MPF) to make it possible to evaluate the effects of the use of the tea by young people. The necessity for it arose when a report by the defunct Federal Narcotics Council (CONFEN), which has been replaced by the National Antidrug Council (CONAD), recommended the prohibition of the use of the tea by those under eighteen years of age. The Juridical Department of the DG [General Directorate of the UDV] sent an inquiry about the matter to the MPF, arguing that the report has no scientific foundation, since nothing has been proved about the supposed harms of the tea for young people. Besides, CONFEN itself had already authorized the use of the Vegetal, since its staff had no record of harm coming from its use by members of the União do Vegetal" (CEBUDV 2003, p. 1).

43. The same publication reported, in a later edition (CEBUDV 2004a, p. 1), that the group managed to collect eighty-nine thousand reais in donations for the project, evidence of the "capacity for mobilization of the Center's members around such an important project as guaranteeing the use of the vegetal by those under 18."

In 2003, Evelyn Doering Xavier da Silveira brought out her psychiatric/medical and psychological research on the neuropsychological evaluation of forty adolescents from the União do Vegetal from three Brazilian cities (São Paulo, Campinas, and Brasília). This research, conducted in collaboration with an international research team, resulted in a series of articles published in a special edition of the Journal of Psychoactive Drugs dedicated entirely to ayahuasca (Doering-Silveira 2003; Da Silveira et al. 2005; Dobkin de Rios et al. 2005; Doering-Silveira et al. 2005a, 2005b).[44]

In brief, these studies verified that adolescents who drank ayahuasca in the context of the UDV were generally similar to the control group in the majority of the tests used in the neuropsychological evaluation. All of the adolescents, both the ayahuasca drinkers and the non-drinking control group, achieved scores within the normal range for every test (Doering-Silveira 2003; Doering-Silveira et al. 2005a). Even when the similarity of the psychological profile of the two groups is taken into account, the researchers showed that compared to the control group, the UDV adolescents had fewer psychiatric symptoms. They appeared to be more responsible, optimistic, respectful and concerned with the well being of others, and had a lower frequency of confrontation and a better quality of life at home (Da Silveira et al. 2005; Dobkin de Rios et al. 2005). Moreover, with regard to the use of psychoactive substances, the UDV adolescents showed results similar to those of the control group, with one notable exception: the UDV group consumed less alcohol (Doering-Silveira et al. 2005b).[45]

The results of the study also showed the influence of gender, socioeconomic factors, and the length of abstinence from ayahuasca consumption. According to the researchers, female adolescent members of the UDV scored significantly higher than their controls on a test that evaluates visual-constructive and visual organization capabilities; these results also provide indirect evidence of the capacity to sustain attention until the completion of the tasks required by the test.

---

44. The primary publications of this project are available on the NEIP site:
   http://www.neip.info/downloads/adoles_udv/arquivos/daSilveira_psPM372pdf,
   http://www.neip.info/downloads/adoles_udv/arquivos/da Silveira_nePM372.pdf,
   http://www.neip.info/downloads/adoles_udv/arquivos/Doering PM372.pdf e
   http://www.neip.info/downloads/adoles_udv/arquivos/de Rios PM372.pdf

45. On the relationship of the UDV to alcohol, see footnote below.

As for the effect of abstinence from ayahuasca in the period leading up to the examination, the only statistically significant difference was found among the male adolescents. Those who abstained from ayahuasca for longer periods of time showed better performance on neuropsychological tests that demand the involvement of multiple cognitive abilities. The male adolescents who abstained for more than forty-five days did significantly better on tests involving attention, concentration, operational (working) memory, and verbal memory than did those who abstained for shorter periods (between twenty and forty-five days). The males who abstained longer showed a tendency toward better visual constructive and visual-spatial organization than did those who abstained for fewer than forty-five days (Doering-Silveira 2003).

Emotional disorders and illnesses caused by damage to brain tissue resulting from the ingestion of neurotoxic substances can influence performance on tasks requiring psychomotor quickness, operational memory, and visual abilities, which are the first skills affected in adverse circumstances. In this context, the hypothesis of the existence of subtle effects of ayahuasca on cognition should be viewed with caution, as other variables, such as socioeconomic factors and membership in a religious group, may be considered protective factors against various mental disorders. According to Doering-Silveira (2003), the design of the study with UDV adolescents does not allow the separate examination of the effects of ayahuasca and the possible positive effect of participation in a religious community.

Keeping in mind that there was a tendency in the control group toward adolescents from more privileged social classes, it is extremely difficult to establish whether the small differences favoring this group in comparison to the adolescents who drank ayahuasca result from socioeconomic disparities, or if such differences should be attributed to a subtle effect of ayahuasca on cognition. Longitudinal studies following changes in the neuropsychological performance of a larger sample of ayahuasca users, over time, may be able to provide more consistent answers (Doering-Silveira 2003).

A qualitative interpretation of the data from the study as a whole has not yet been published. The great importance of these articles lies in their status as the first to evaluate the effects of ayahuasca in adolescents who consume it regularly, a subject that will surely take on fundamental importance in the public debate about the use of ayahuasca by children and adolescents, particularly outside Brazil.[46]

Other studies in Brazil and overseas

In 1998, Eliseu Labigalini, Jr., one of the Brazilian researchers who participated in the Hoasca Project, studied four individuals who suffered from severe alcohol dependency for his master's degree in mental health. The four appeared, several months after beginning to participate in União do Vegetal rituals, to have changed their behavior completely (Labigalini Jr. 1998).[47] Labigalini presented his findings, with collaborators, in international journals (Labigalini Jr. and Dunn 1995; Miranda et al. 1995). Although these studies lacked control groups, placebos, and double-blind design,[48] they indicate a possible effect of ayahuasca in the treatment of cases of dependence on alcohol and other drugs which should be more fully explored.

Also working in the area of mental health, Paulo Cesar Ribeiro Barbosa has conducted research from the perspective of cultural psychiatry, working with twenty-eight individuals who drank ayahuasca for the first time in ceremonies of Santo Daime and the União do Vegetal in São Paulo and Campinas (São Paulo state) (Barbosa, 2001, 2003, 2004).[49] The results were published in Brazil (Barbosa and Dalgalarrondo 2003) and overseas (Barbosa et al. 2005). The research evaluated the states of consciousness induced by the ingestion of ayahuasca in people from urban contexts who had no prior experi-

---

46. In earlier reports issued in Brazil by the old Federal Narcotics Council (CONFEN, now replaced by the National Antidrug Council, CONAD), there was a recommendation that minors under eighteen not use ayahuasca (CONAD 2002; Labate 2005). In the resolution dated November 4th, 2004 (CONAD 2004), which is the current one, and in the final report of the Multidisciplinary Task Force on ayahuasca (CONAD 2006), CONAD retracts this contraindication and gives parents the right to decide, thus placing the issue within the domain of "family rights," that is, in the domain of the private rather than the responsibility of the state. In other words, the decision whether to give ayahuasca to one's children is analogous to all those related to raising and educating children, since just as liberty of religion is an individual's constitutional right, so too is the liberty to teach and the right to raise one's own children as one sees fit. In practice, however, the issue continues to create conflict, for example in cases of separated couples who disagree on this question, and sometimes even generates legal disputes.

47. The work of Labigalini Jr. is available on the UDV site: http://www.udv.org.br/portugues/downloads/13.rtf. Resultados semelhantes foram encontrados no Projeto Hoasca (Grob et al. 1996a, 1996b, 2004).

48. The double-blind design refers to a methodology in which neither the volunteers nor the researchers directly involved in the project know the nature of the drug administered (for example, whether it is a placebo or ayahuasca). The use of the placebo—an innocuous substance—is more common in studies that evaluate the acute effects of consuming a given substance. Experimental studies in which psychological and neuropsychological results of chronic consumption are assessed do not lend themselves to placebo design, since it would entail a research plan in which one would administer a certain substance to one group and a placebo to another, which in practice is very difficult to carry out. In addition, the long-term administration of a substance whose chronic effects are not known would not be ethical.

49. The study by Barbosa is available on the NEIP site: http://www.neip.info/downloads/paulo_tse/tese_paulo.pdf.

ence with the drink. The twenty-eight study participants (nineteen from the Santo Daime denomination Cefluris in São Paulo and nine from the União do Vegetal in Campinas and São Paulo) were evaluated between zero and seven days before, and zero to seven days after their first experience with ayahuasca. All the evaluations were initiated with the administration of a structured clinical interview, followed by a socio-demographic questionnaire, and were completed with qualitative interviews.

In this study Barbosa found two principal experiential patterns induced by the ingestion of ayahuasca, which he terms "serenity" and "power." "Serenity" was characterized by silencing, calming, and soothing; "power" was marked by a "numinous" tone.[50] Barbosa recognized in these mental states, induced with ayahuasca, radical structural and qualitative changes relative to the normal state of consciousness. In addition, the author observed a partial autonomy of these mental states in relation to the religious circumstances, indicating that ayahuasca may produce certain effects independent of the context, motives, and expectations of the experience: thus the experiences came as a surprise to many participants. No medical or psychotherapeutic interventions were registered.

This study demonstrated a drastic reduction in psychiatric symptoms for some subjects, with a general improvement in emotional state and shifts toward more passive / assertive attitudes. However, Barbosa affirms that because such changes decline gradually in the days following the use of ayahuasca, they can be characterized as remnants of the experience, rather than permanent changes. In addition, the author mentions a case of psychiatric institutionalization, following the ritual use of ayahuasca, of an individual who was not part of the study. In the initial psychiatric evaluation, the patient presented psychotic symptoms (discourse with delirious content, visual hallucinations, unmotivated laughter, bizarre behavior, and loose association of ideas, among others), which were controlled after five days of internment. According to Barbosa, this subject had a predisposition to psychotic symptoms, and at age fourteen had been hospitalized with psychiatric symptoms including paranoid ideation and auditory hallucinations; in addition, approximately three weeks

---

50. The term "numinous" was coined by the theologian Rudolf Otto in his book The Sacred, written in 1917, and connotes the living, original, and non-rational element present in all religions, thesacred—direct contact with a transcendent "Higher Power" (cf. Carvalho 2005, pp. 54-58).

before he drank ayahuasca a friend had hung himself, an event that affected him significantly. Lastly, the subject's work life was overloaded at the time of hospitalization. In Barbosa's interpretation, while ayahuasca may have played a role in the episode, it was not the only or even the most important factor in triggering a psychotic reaction, to which the subject, given his prior hospitalization, was particularly vulnerable. Besides this tendency, there were professional demands and personal losses that appear to be the principal determinants of the episode. Still, the case suggests that individuals with a predisposition to psychosis should not drink ayahuasca. Such precautions are already well known to apply to hallucinogens in general (Grinspoon and Bakalar 1981; Strassman 1984; Masters and Houston 2000; Grof 2001), and according to our field observations have been observed, generally speaking, in both Santo Daime and the União do Vegetal.[51]

Barbosa's approach is innovative in that it assesses first-time users of ayahuasca, rather than relying on retrospective reports by long-time, "converted" drinkers, whose narratives may be mediated by the reinterpretation of their past through the lens of the group's religious categories (as in the case of the Hoasca Project, for example). The results of his study, however, should not be understood as representative of ayahuasca experiences in the context of these religions, but should be seen for what they are: evaluations of initial experiences with the drink. Even so, and despite the lack of a control group and the other methodological tools discussed above, the eventual replication of these data may provide interesting views on the therapeutic potential of ayahuasca. Barbosa is currently developing a longitudinal assessment of the volunteers from his 2001 study (Barbosa 2006).

It is noteworthy that, despite the existence of thousands of ayahuasca drinkers in Brazil,[52] where the ritual-religious use of this substance is permitted, and given the expansion of this practice to cities in Europe, North America, and

---

51. According to our fieldwork in various parts of Brazil over the last ten years, preliminary interviews prior to participation in rituals with ayahuasca aimed at selecting and informing participants do tend to take place in most of the groups but not in all of them. It is worth recalling that the Brazilian government counsels against administering ayahuasca to "carriers of mental deficiencies," to people with a "history of mental illness," and suggests conducting interviews beforehand (CONAD 2002, 2006). Of the groups studied, the UDV appears to be the most rigorous, and has even produced studies on the theme (Lima 1996-1997; Lima et al. 1998, 2002). In the case of Santo Daime, there is a standard interview available on the group's official site at http://www.santodaime.org/doutrina/cura/iniciantes.htm. The practice of preliminary interviews should be encouraged with the aim of protecting certain psychologically vulnerable individuals as well as the group itself.

52. For a tentative estimate of the total number of users of ayahuasca in Brazil see "Bibliographical overview of the ayahuasca religions" in this volume.

Japan (among other countries), Barbosa's studies are, thus far, the only ones to examine the acute effects of ayahuasca on the psychiatric symptoms of people drinking it for the first time. Other studies of this type should be conducted to increase our understanding of the acute effects of ayahuasca on psychiatric symptoms.

In 2006, Rafael Guimarães do Santos defended a master's thesis on the relationship between ayahuasca and states of anxiety, depression, and panic.[53] The study was carried out in Brasília with nine members of Santo Daime who had drunk ayahuasca for at least ten consecutive years. The author found an attenuation of levels of depression and panic approximately one hour after the ingestion of ayahuasca. The study's sample size was small, and its reliance on regular drinkers of ayahuasca may have influenced the results, since these individuals already had experience with the brew. Furthermore, the use of healthy volunteers may limit the study's clinical findings. Nevertheless, replication of these results should be sought in order to explore the therapeutic possibilities of ayahuasca with better methodology.

Between the Hoasca Project and the studies of UDV adolescents came a series of other research projects, both Brazilian and international, though none of these has produced the impact of the first two. Many of the studies of toxicology, psychology, and psychiatry (for example, Labigalini Jr. 1998; Cazenave 2000; Camargo 2003; Costa et al. 2005) lack the necessary methodological rigor, which has contributed to the scant notice they have received. The Brazilian studies, having been published in Portuguese, have failed to gain recognition within the international scientific community. Finally, these studies are little known in part because of the field's incipient status. Nonetheless, there currently seems to be a boom in Brazilian and international studies of ayahuasca, especially in the field of anthropology.[54]

An emerging field: Studies in Spain

In 2006, Rafael Guimarães dos Santos began a PhD study in Barcelona, Spain on the human pharmacology of ayahuasca. This project, unlike those already cited, is being carried out in a laboratory setting instead of in the context of

---

53. The study by Santos is available at the NEIP site: http://www.neip.info/ downloads/rafael/tese_rafa.pdf. This study has recently been accepted for publication in a well-regarded international publication (Santos et al. 2007).

54 See "Bibliographical overview of the ayahuasca religions" and "Bibliography of the ayahuasca religions" in this volume.

the ayahuasca religions (Santos 2007a). It is mentioned here because it is part of the investigations conducted by the team of doctors Jordi Riba and Manuel J. Barbanoj of the Universidad Autónoma de Barcelona. The results of these, the most recent and most rigorous studies of ayahuasca's human pharmacology, have been coming out since 2001 in respected international journals (Riba et al. 2001a, 2001b, 2002a, 2002b; Yritia et al. 2002; Riba et al. 2003, 2004, 2006).[55] The ayahuasca used in these studies, however, is a freeze-dried preparation.[56]

Riba and colleagues have used double blind, placebo-controlled experimental designs to evaluate the acute effects of ayahuasca in healthy volunteers with previous experience with hallucinogens. With this methodology, the researchers sought to avoid, as much as possible, interference of volunteers' and researchers' expectations with the results. Furthermore, such studies are conducted with controlled doses of ayahuasca in which the quantity of alkaloids is known. In this way, the results can be attributed to specific doses of the ayahuasca alkaloids, although these doses may not necessarily be replicated precisely in the consumption context of the ayahuasca religions.

The studies by Riba published thus far have evaluated ayahuasca's acute subjective effects, as well as its tolerability and cardiovascular effects, its electroencephalogram profile, the distribution of the alkaloids and their metabolites, the brain areas involved, etc. In future studies, these researchers will examine the brain receptors involved in ayahuasca's effects and the substance's acute immunological effects, among other dimensions (for a general analysis of the results already published, see Riba and Barbanoj [2005]).

For these reasons, the studies by Jordi Riba and colleagues, with their modern experimental designs, are currently the ones that can answer questions about the effects of ayahuasca in human beings with greatest confidence from the biomedical and pharmacological points of view. Still, it is worth pointing out

---

55. Some of these articles can be found in the PhD thesis by Jordi Riba (2003), available at the site of the Multidisciplinary Association for Psychedelic Studies (MAPS): http://www.maps.org/research/ayahuasca/jriba_thesis.pdf.

56. In brief, liophylization can be understood as a process of dehydration (freeze-drying). In the case of ayahuasca this means that the water was removed from the preparation. The final product is a powder, which is then stored in capsules. It is not a synthetic ayahuasca. This process permits better conservation of the alkaloids and allows the ayahuasca to be placed in capsules identical to those used from the placebo (a pharmacologically inactive substance). In practice, this means that both the ayahuasca and the placebo are administered in the same way, making their identification by both volunteers and investigators more difficult. For details, see Riba and Barbanoj (2005).

that the fact of ayahuasca being administered in capsules and in a hospital environment influences the experience of participants in these studies. This is particularly true if volunteers understand the environment as "cold" or even threatening, or interpret it as "fake" in contrast to a ritual or natural setting which may be considered more "authentic"—just as, at the other end of the spectrum, the context of consuming ayahuasca in a ritual may likewise be perceived, by certain individuals, as "authoritarian," "dogmatic," or frightening, and thus determine the course of their experience.

Obviously, the hospital context is distinct from these other environments and is not exempt from its own influences on the nature of the experience, beginning with the very process of selection and interaction in choosing participants for the study, which creates certain expectations. Many other factors may influence the results of the experience and the collection of data. To cite just a few examples, there are reports of participants who learn to detect whether they are given a placebo or ayahuasca because of the taste of the substance (Dávila 2007);[57] in addition, in certain experiments it is inadvisable to vomit during the "session," since vomiting can alter the concentration of alkaloids, whose precise levels in the blood are important to the study's objectives—a fact that imposes certain pressures on the experience that are particular to this context.

From a scientific point of view, the advantage of administering ayahuasca in capsules in a hospital context is that such studies are considered more controlled, since the levels of alkaloids administered are known and it is possible to do tests that would be impossible or very limited in a ritual context (for example, measuring cardiovascular changes in a dancing ritual or conducting electroencephalogram tests during the ceremonial preparation of ayahuasca). There is no question of trying to duplicate ritual and religious contexts in a hospital environment, yet it is interesting to note that Jordi Riba and colleagues have demonstrated changes in certain variables similar to those found

---

57. It should be made clear, however, that this was speculation on the part of the volunteer since, after all, he did not have access to data about the substance ingested. In any case, in some cases it may be possible for volunteers to identify ayahuasca by its effects. In other cases, as in our fieldwork observation of rituals of the ayahuasca religions, the individual may not feel any effect, even after ingesting more than one dose of ayahuasca. Furthermore, according to Santos (2006; Santos et al. 2007), experienced ayahuasqueiros with at least ten years of continuous consumption of ayahuasca ingested a placebo preparation and one of the volunteers, on the day the placebo was administered (when everyone, including him, knew what was being administered) thought he had ingested an ayahuasca preparation, demonstrating the power of suggestion.

in studies conducted within the ayahuasca religions, such as the Hoasca Project; there are also reports of similar experiences in other kinds of hospital contexts.[58]

Since the goal of this chapter is to review studies on the basis of criteria established through the history of scientific study of hallucinogens (which, in large part, were and still are conducted in hospital and clinical settings—see Grinspoon and Bakalar 1981; Strassman and Qualls 1994; Strassman et al. 1994, 1996; Masters and Houston 2000; Grof 2001; Jansen 2004), we cannot neglect to mention, if only in passing, a problem underlying studies done in environments designed for the experience. The problem involves the representation of the clinical context as a domain of pure objectivity, free of cultural influences, and of the results of studies conducted in it as "biological" findings that are therefore universally valid. In fact, the clinic and the laboratory have their own culturally and socially determined logic, with its own rules, jargon, esthetic, interpretive categories, and even its own little rituals. As we have said, this text is not intended to develop this kind of analysis, but is restricted to a panoramic description of existing studies. Nevertheless, the discussion of the laboratory context itself is one possible path to anthropological reflection on scientific studies.

As can be observed, a large proportion of the studies conducted on the ayahuasca religions have been done within the UDV. The UDV seems to be a good "laboratory" insofar as the institution discourages the use of other psychoactive substances (licit or illicit),[59] thus providing an ostensibly good sample of "pure" users of ayahuasca, at least in the sense of the influence of

---

58. Strassman and colleagues conducted experiments with N,N-dimethyltryptamine (DMT), a substance found in samples of ayahuasca, in a hospital context. The hormonal changes and subjective effects reported in the studies by Strassman—increases in the levels of cortisol, prolactin, and growth hormone, as well as reports of visions—were also found in the Hoasca Project studies (Strassman and Qualls 1994; Strassman et al. 1994, 1996; Callaway et al. 1999).

59. Article 17 of the Internal Regulations of CEBUDV, which is read in each "scale session" (every two weeks), states that "any member found in a state of visible drunkenness will be warned by the General Administration and, in case of reincidence, will be punished for disobedience." While the article does not explicitly prohibit the use of alcoholic beverages, but instead forbids the ingestion of a quantity that leads a member to a "visible state of drunkenness," in practice the use of alcohol and also tobacco are not permitted within the UDV, since it is taught that human beings, to "evolve spiritually," must quit these habits. As for illicit substances, there is no law in the UDV that directly prohibits them, but it can be said that they are prohibited, as the UDV preaches rigorous fulfillment of the laws of the country. In our fieldwork we frequently observed testimony of the sort: "before I came here I was lost, addicted, but now I've found my path." At the same time, we saw a certain tolerance toward those who have frequented the group for a short time and are perhaps still "freeing themselves from vice." The members of the upper echelons of the hierarchy, on the other hand, may not smoke, drink, or consume other "drugs." It is worth remembering, however, that rules do not tend to be followed to the letter. Other ayahuasca religions also disapprove of the use of alcohol, tobacco, and "drugs," as is the case of the Barquinha. For more information about the attitudes of Santo Daime, the UDV, and the Barquinha toward "drugs," see Goulart (2004).

Man cooking Daime – Céu do Mapiá, Amazônia – 2007
Crédito: Photo by Andréa D'Amato

the use of other substances. In addition, the group has an excellent organizational structure spread throughout Brazil and includes some ethnic and social variation. Finally, because it is fairly cohesive, centralized, and hierarchical, and because it values science and the affirmation of Mestre Gabriel, the religion's founder, that "hoasca is inoffensive to health," it has provided enthusiastic volunteers for these studies.

The UDV and the field of biomedical studies of ayahuasca

In addition to having been the "object" of these studies, the UDV, it is important to note, has played a central role in current scientific work on ayahuasca, insofar as the two principal studies involving human subjects (the Hoasca Project and the adolescent study) came about due to wide-ranging and intensive efforts by its Medical-Scientific Department (DEMEC) (CEBUDV 2003, 2004a, 2004b; also see above).

In this way, it becomes clear that the UDV is the ayahuasca-using group with the greatest interest in legitimating the use of ayahuasca from a medical-scientific perspective. In fact, DEMEC was created for this very purpose. According to the UDV website, when the vine *Banisteriopsis caapi* was temporarily placed on a list of controlled substances in 1985 by the Health Ministry's Division of Medicines (DIMED)—a function now performed by ANVISA, the National Health Surveillance Agency—members of the UDV who work in the health professions organized themselves to collect scientific information about ayahuasca. In 1986 the UDV created the Center for Medical Studies, now called DEMEC. Among other functions, this department evaluates and oversees scientific studies focused on the UDV. DEMEC is also charged with representing the UDV in interactions with legal and scientific institutions, and has supported the research cited above.[60]

---

60. According to the UDV site, among DEMEC's other functions "it is the organ of interaction between the Center and the academic world, and the facilitating agent of the institutional conditions necessary for scientists to conduct research"; it represents the UDV "before legal and scientific authorities" and strives "for the fulfillment of the UDV Health Professionals Charter of Principles." According to this charter, "health professional members of [the UDV]" declare that they "accept the União do Vegetal Center of Medical Studies as the supervisory organ of scientific studies and experiments relevant to Hoasca carried out by members" of the UDV; "experiments and scientific studies supervised by the Center of Medical Studies will be guided by the absence of preconceptions, by appropriateness, and by decorum, and the researchers will refrain from divulging the results, which will only be done at the appropriate time by the Center of Medical Studies"; "they will not make professional use of Hoasca, in or out of the context of [the UDV], unless there is scientific proof of its therapeutic qualities"; however, the charter of principles "may be reevaluated should there be official, scientific proof of the therapeutic efficacy of Hoasca tea" (http://www.udv.org.br/portugues/texto/udv-versao-texto.doc). On general aspects of DEMEC and the UDV, see also Goulart (2004) and Labate (2004, 2005).

It was not by chance that the UDVs conceived of, encouraged, and partially funded such studies. In general, as with other psychoactive substances, biomedical studies of ayahuasca tend to carry a decisive legitimacy in public debate on the topic, and to meet the demands of the centers of power in contemporary society. It may be no exaggeration to affirm that there is, generally speaking, a nearly exclusive recognition of the biomedical studies as the only consensually relevant and consistent ones in the evaluation of the risks and harms of the use of ayahuasca. Nevertheless, in the global debate on the matter, biomedical discourse assumes a different role in different countries.[61] It should be emphasized that this attitude of the UDV is the result not only of a political strategy, but also, in part, of factors in its religious cosmology, in which a particular concept of "science," distinct from modern western science, is to be found.[62] Be that as it may, such an institutional orientation of the UDV has spurred the inauguration of a new phase in studies of ayahuasca.

Heath sciences, social sciences and users: crossed perspectives

From the point of view of hegemonic scientific conceptions, the methodology described in some of the studies cited above, especially those of Jordi Riba and colleagues, represents the best way to organize and understand the biochemical and physiological aspects of ayahuasca, and, in some cases, its therapeutic efficacy. However praiseworthy the efforts of these scientists may be, joined as they are by sophisticated transnational research networks, it should be remembered that the consumers of these substances are, without the shadow of a doubt, those most empirically familiar with the effects, virtues, and problems stemming from their use.

In the cases of Santo Daime and the União do Vegetal, our fieldwork has shown that the leaders of these groups have developed expertise in administering ayahuasca differently depending on the context (the type of ritual)

---

61. There exist, however, studies of the therapeutic potential of ayahuasca from other points of view, such as the work of Peláez (1994), Labate (2004), and Rose (2005), which discuss the "native" views of the relationship between ayahuasca and healing. In this sense, it is important to note that there exists a tradition in medical anthropology that highlights the relativity of notions such as "healing," "health," "illness," and "therapy," emphasizing that there are other ways, besides the biomedical point of view, of understanding these concepts.

62. In the UDV, "science" is valued as a sort of true knowledge whose nature is also spiritual and which is associated with the figure of Solomon from the Judeo-Christian biblical tradition. On the role of Solomon in the UDV, the UDV concept of "scientification" and their relationship to Kardecist spiritism and to other traditions, such as the Judeo-Christian, see Goulart (2004); for an overview of the role of "science" in the UDV, see Labate (2004, 2005). More studies are needed to understand how and to what degree aspects like the science of Solomon and initiatives of a secular scientific character like those of DEMEC are linked in the UDV.

and the individual: for example, during pregnancy and breastfeeding; to children and adolescents; and to those with psychological disorders, certain health problems, or who are taking particular medications, etc. Thus, the dosage and regularity of ayahuasca consumption vary considerably. Perhaps an example of this will be useful. The specialized literature suggests that the use of certain anti-depressive medications with ayahuasca may result in adverse reactions, some of which might be fatal. Such reactions may also be produced by the combination of ayahuasca with pro-serotonergic substances in general, like various kinds of antidepressants, the amino acid tryptophan, and "ecstasy" (MDMA). Other adverse reactions can result from the combination of ayahuasca with certain foods or dietary supplements (Callaway 1994; Lima 1996-1997; Callaway and Grob, 1998; Santos 2007b). In our field research, we observed that a good part of the Brazilian ayahuasca groups adopted recommendations of caution in relation to the interaction of ayahuasca with certain serotonin selective reuptake inhibitors, such as fluoxetine (Prozac).

We also observed, however, that in practice through the years, both Santo Daime and the UDV relaxed restrictions to these medications on certain occasions because of their empirical observation of the possibility of ingesting ayahuasca with them without great harm. Still, such administration is done very cautiously, depending on a series of factors such as: the type of personality, the dose of the medication, the length of treatment with it, prior history of psychological disorder, regularity of attendance at rituals, general state of physical and spiritual health, concentration of the ayahuasca, etc. Here there seems to be an interesting combination of medical and folk knowledge at work, one which has not been reported in the literature and which is subject to a certain taboo in some contexts.

In addition, we have noted that both groups have collected a series of reports of treatment or "cure"[63] of various illnesses, from depression to cancer, through participation in the rituals, with frequent accounts of "cures" of "chemical dependence." By the same token, there is ample empirical experience of persons who have had various problems with the use of ayahuasca

---

63. The Daimista concept of healing is broader than the biomedical one. Explanations of the causes of illness encompass dimensions that are not emphasized in biomedicine, and which allow healing to fulfill functions beyond the remission of symptoms. Great emphasis is placed on spirituality, so that healing is understood fundamentally as "spiritual healing" (Rose 2005).

(Lima et al. 1998, 2002). In the case of Santo Daime, there is even a term used to designate persons who experience problems drinking ayahuasca: the "rodados."[64]

All of this knowledge remains in large measure unrecognized and is often little systematized within the groups themselves. The knowledge accumulated by DEMEC has not yet been published,[65] perhaps because of fear that favorable results may be interpreted as a biased defense of ayahuasca and the negative results be taken as evidence of its "dangers," or perhaps merely because the group wants to solidify its own research more before making its results known. It is possible, however, that in the future DEMEC may publish a book containing the results of its long years of observation. The UDV is thus a fascinating case within the world of ayahuasca consumption because of its combination of the accumulation of broad empirical experience with monitoring by specialists in medical fields.[66]

The studies done by DEMEC and the testimonies of the consumers themselves have much to contribute to scientific knowledge, and especially to the neurosciences: more elaborate qualitative responses, data on the effects of ayahuasca in children, adolescents, and pregnant and breastfeeding women; plus techniques and suggestions for the administration of ayahuasca with reduced risk, knowledge about the doses used in special circumstances (such as those of people in psychiatric treatment), etc. From this perspective, the health sciences should be more open to the knowledge held by these groups, and the groups to scientific knowledge.

---

64. The term "rodar" [Port., "to turn," "to spin"] may be a reappropriation of a term used in Umbanda, in which "rodar" means to be "incorporated," when a medium's body spins in a dance that may be either leisurely turns or violent, out-of-control whirling (Dias 2007).

65. DEMEC doctors have authored or co-authored articles in the national (Brazilian) and international literature, but DEMEC has not yet produced its own publications. Some of the UDV health professionals who have published in this area are: Glacus de Souza Brito, Edison Saraiva Neves, Francisco Assis de Sousa Lima and, to a lesser degree, Otávio Castello de Campos Pereira, Mauro Bilharinho Naves, Júlia Motta, and José Carlos Veras Di Migueli. Glacus Brito works with orthomolecular medicine, ozone therapy, and homeopathy; Edison Saraiva Neves works with Unicist homeopathy, medical nutrology, orthomolecular medicine, and ozone therapy; Francisco Assis de Sousa Lima is a psychiatrist; Otávio Castello de Campos Pereira is a geriatric specialist; Mauro Bilharinho Naves is a psychiatrist; Júlia Motta is a psychologist and psychotherapist; José Carlos Veras Di Migueli is a psychiatrist and psychoanalyst. Some of the publications are: Callaway et al. 1994; Callaway et al. 1996; Grob et al. 1996a, 1996b; Lima, 1996-1997; Lima et al. 1998; Callaway et al. 1999; Lima et al. 2002; Andrade et al. 2004; Brito, 2004; Grob et al. 2004; Callaway et al. 2005.

66. This kind of combination of science, religion, and local knowledge is not exclusive to the União do Vegetal, but is related to the advent and expansion of the ayahuasca religions. An interesting example of this interaction is the Cefluris church Céu da Mantiqueira, in Camaducaia (Minas Gerais state), studied by Isabel Santana de Rose (2005). This church, at the time of research, had among its members several psychiatrists and other health professionals, including some who also study ayahuasca and psychoactive substances in general, as in the case of Eliseu Labigalini Jr. (Labigalini Jr. and Dunn, 1995; Labigalini Jr. and Rodrigues, 1997; Labigalini Jr. 1998) and Lúcio Ribeiro Rodrigues (2006). The church was known in Cefluris as a "center for healing." In the "line of Padrinho Sebastião," however, there is not a very great preoccupation with the role of scientific knowledge in the rituals and outside them such as exists in the UDV.

Pharmacology and medicine, for example, may be able to offer users and researchers new ways of looking at the phenomena of consuming hallucinogens in general and ayahuasca in particular by providing data on the biochemistry of mystico-religious experiences, on the occasional risks associated with the consumption of these substances, and on the relationship between neurochemical changes and the experience of different states of consciousness.[67] Studies in the social sciences, as well as folk wisdom, provide important information about the world of consumption of these substances which, as we argued at the outset, cannot be considered apart from the contexts in which their use occurs.

Future perspectives

We have word of more studies in progress in Spain on the biomedical and psychological aspects of ayahuasca among Santo Daime and Barquinha groups in Acre and Amazonas (José Carlos Bouso, personal communication, June 2007). These are longitudinal studies of about sixty members of Cefluris/Santo Daime and sixty members of the Barquinha which compare them with a control group of equal size, and aim to evaluate the long-term effects of ayahuasca consumption, including those on personality, neuropsychological function, general health, psychosocial well-being, and spiritual values. The exams were conducted and then reapplied eight months later with the objective of analyzing the stability of the results. The researchers are planning a second evaluation one year after the first and a third, one year after the second in order to assess the stability of the scores. These studies may reveal significant findings.

---

[67]. It should be remembered, however, that many scientific studies may run afoul of religious sensibilities. The methodology of these studies may be criticized as "cold," "reductionist," and "decontextualizing," among other things. It was in this vein, for example, that the leader of a small group from Campinas (São Paulo state), Andrade Neto (see "Bibliographical overview of the ayahuasca religions" in this volume for more information about this group), speaks of projects on the Vegetal: "I consider them futile to the point of nonsense... [T]he act of dissecting the two plants and analyzing their chemical effects on the human organism belongs to those who have not understood anything about their importance, and who haven't been able to understand the purpose of this powerful tea... [T]o insist that people who have never drunk the tea publish reports on the subject implies the assumption that this same tea falls within the reach of human judgment, and therefore it also implies the negation of its superiority before humans..." (Rodrigues 1998, pp. 86-7). Carrying out these studies may also be considered an arbitrary attitude, and their legitimacy problematized for political reasons. Thus the Colombian Union of Indigenous Yagé Doctors [Unión de Médicos Indígenas Yageceros de Colombia (Umiyac)] declared in the annals of the meeting, which brought together 140 indigenous people from six different ethnic groups in Yurayako, Caquetá, in 1999: "We denounce the experiments with yagé and medicinal plants carried out by many anthropologists, botanists, doctors, and other scientists without taking into account our ancestral knowledge and our collective intellectual property rights" (Umiyac 1999, p. 110). The project of making "science" and "native knowledge" dialogue with one another includes relations of mutual apprenticeship, but also of tension, as well as exchanges which are not always egalitarian, but frequently hierarchical.

One question that is still little studied in this field is that of the therapeutic potential of ayahuasca or of some of its alkaloids, particularly from a biomedical perspective. The studies examined in this chapter do not, with few exceptions, deal directly with therapeutic issues. In some of these studies the results indicate possible therapeutic effects, but as far as we can tell, there have as yet been no clinical, controlled studies on these potentials. Some preliminary research points to ayahuasca's therapeutic potential in, for example, the treatment of depression, anxiety, panic, or substance abuse (Miranda et al. 1995; Grob et al. 1996a, 1996b; Labigalini Jr. 1998; McKenna et al. 1998; Grob et al. 2004; McKenna 2004; Santos 2006; Santos et al. 2007). These studies are limited, however, and indicate the need for further research.

It is therefore important to remember that the National Anti-drugs Council (CONAD) established a Multidisciplinary Task Force (GMT) "to study and monitor the religious use of ayahuasca, as well as to research its therapeutic uses through experimental means" CONAD 2004). The GMT stated, in its final report (CONAD 2006, p. 10):

> Any practice that involves the use of Ayahuasca for strictly therapeutic ends, whether of that substance alone, or in combination with other substances or therapeutic practices, should be prohibited, until such time as its efficacy is shown by means of scientific research conducted by research centers affiliated with academic institutions according to scientific methodology. Thus, the recognition of the legitimacy of therapeutic use of Ayahuasca will only take place following the conclusion of studies that demonstrate it.

The same report suggests that scientific study be stimulated in the areas of pharmacology, biochemistry, clinical practice, psychology, anthropology, and sociology, and that CONAD, beginning in 2007, should finance research on the use and effects of ayahuasca. It is possible, therefore, that we are witnessing the birth of a new generation of biomedical research in Brazil. As regards the role of Brazil in the international field of pharmacological, psychological, and psychiatric research on ayahuasca, it can be said that, despite the origin of these religions in that country, it has produced little literature on them; the few projects done in this area, however, have received significant momentum from Brazil.

Moreover, future biomedical studies of ayahuasca may, by elucidating its mechanisms of action in the human nervous system, expand our knowledge of the mind-brain relationship, of the biochemistry and physiology of perception, emotion, and cognition, and of the formation of consciousness.

At the same time, this short review of the pharmacological, psychiatric, and psychological studies of the ayahuasca religions reveals the fundamental necessity of anthropological research on the formation of this field of inquiry in order to understand the conditions of its production and the multiple factors that interact in its realization and the dissemination of its results. Several questions are relevant in this sense. What is the relationship between "native" researchers and "scientific" researchers in the production of these studies? Can the two categories be clearly distinguished? Are there researchers who belong to both categories? How have these studies been appropriated in the public arena, and how may they be appropriated in the future? How do they influence legal processes involving the legalization of ayahuasca use in various countries?[68] Is the UDV litigation in the United States stimulating new research on ayahuasca, or is the group's leadership blocking new studies from being carried out?

Many other questions might be asked as well, and we point to just a few to encourage the development of new research. Might there be a division of labor forming between North Americans (who have embraced the study of the UDV) and Spaniards (who are beginning to study Santo Daime and the Barquinha)? How were such endeavors as the Hoasca Project and the adolescent study envisioned and made possible, including from a financial point

---

68. An incipient example may be found in the case of the legal case that the UDV has been waging for several years in the USA. One of the arguments used by the government of the United States in an attempt to delegitimize the testimony presented by the UDV was that "most of" the UDV's witnesses were "either members of UDV and thus prospective users of hoasca themselves or conducted research funded by the head of UDV, respondent Bronfman" (cited in Meyer 2006, p. 30) – referring here mainly to the Hoasca Project, which received donations from the Mestre Representante of the UDV in the USA, Jeffrey Bronfman (who also donated funds through the Aurora Foundation, of which he is president, for other studies, such as the chemical analyses of the vine *Banisteriopsis caapi* and the leaf of the shrub *Psychotria viridis* done by Callaway et al. (2005), as indicated by a note of acknowledgment in the article, on page 145). Even though this is a rather predictable strategy on the part of the North American government in its tireless "war on drugs," we are still far from understanding the multiple variables involved in the creation of the broad transnational networks of researchers involved in these projects. It is certain, however, that any researcher writing about ayahuasca today knows that his or her work may be appropriated in judicial disputes. Thus Callaway et al. (2006), after strongly criticizing the study by Sklerov et al. (2005), conclude by demanding responsibility on the part of the researchers, since incorrect aspects of the supposed case of toxicity involving the use of ayahuasca reported by Sklerov et al. "could influence future legal decisions on this matter in the United States and in other countries" (2006, p. 407).

of view? What is the relationship between the UDV's adolescent study and CONAD's decisions on the use of ayahuasca by minors? How have the concerns and anxieties of the health and law professionals in the UDV influenced the political and even the theological paths the institution has taken? How do "medical" and "caboclo" ideas about the Vegetal exist together within the UDV? Do DEMEC researchers tend to hold positions of spiritual authority (such as Mestres and Conselheiros) within the UDV? What conception of "science" predominates in the UDV? How does DEMEC handle internal investigations of psychiatric incidents with ayahuasca?

It is worth remembering that a number of these questions can be asked, in an analogous way, about Santo Daime, the Barquinha, and the new urban groups that have formed in Brazil's metropolitan areas,[69] many of which, incidentally, are led by psychiatrists or psychologists.

Based on these reflections it becomes clear that research on substances like ayahuasca has a lot to gain from interdisciplinary study involving native knowledge, anthropology, sociology, psychology, pharmacology and medicine, in addition to law, religious studies, and disciplines relevant to public health. As a bio-psycho-social phenomenon, only a similarly diverse perspective can provide an adequate view of ayahuasca. At the same time, it is necessary to view biomedical research through the anthropology of science, seeking to contextualize the production of scientific knowledge.

Finally, a short historical comment: in the United States and in Europe between the 1940s and the 1970s, research was done on hallucinogens as much by "alternative" researchers as by those affiliated with more orthodox academic institutions. In the United States of the 1950s and 1960s researchers such as Aldous Huxley, Timothy Leary, and Richard Alpert began a new tradition of studies outside the academic mainstream and thus inaugurated a new epistemological field. By the end of the 1960s worldwide research with hallucinogens underwent a rapid constriction. Today, thirty years later, we see a rebirth of this area of research.[70]

---

69. For a discussion of urban neo-ayahuasqueiros, see Labate (2004).

70. For more information, see the website of the Multidisciplinary Association for Psychedelic Studies (MAPS): www.maps.org, especially the page on research with hallucinogenic substances conducted in various parts of the world: www.maps.org/research.

It bears pointing out in this context that although some of the pharmacological and biomedical researchers mentioned here are associated with important universities in the United States, Canada, Europe, and Brazil (such as Dr. Charles Grob of the University of California at Los Angeles, Dr. Dennis McKenna of the British Columbia Institute of Technology, Dr. Jace Callaway of the University of Kuopio, Dr. Jordi Riba of the Universitat Autònoma de Barcelona, and Dr. Dartiu Xavier da Silveira, of the Universidade Federal de São Paulo), the majority of the studies mentioned here were carried out by researchers who are outside of the international academic mainstream (e.g., Quinlan 2001; Cougar 2005; Sulla 2005), and their results appeared in lesser known periodicals, as is the case with the Journal of Psychoactive Drugs, where the UDV adolescent study appeared. Thus, despite the current rebirth of hallucinogen research, this field of study still finds itself marginalized in a certain way. We hope that the research presented here, as well as these incipient reflections on this field of study, may serve as a stimulus for researchers in the biomedical and in the social sciences to try to carry out further investigations of the ayahuasca religions.

Bibliographical references

ANDRADE, E.N.; BRITO, G.S.; ANDRADE, E.O.; NEVES, E.S.; MCKENNA, D.; CAVALCANTE, J.W.; OKIMURA, L.; GROB, C.; CALLAWAY, J.C. (2004). "Farmacologia humana da hoasca: estudos clínicos (avaliação clínica comparativa entre usuários do chá hoasca por longo prazo e controles – avaliação fisiológica dos efeitos agudos pós-ingestão do chá hoasca)," in: LABATE, B.C.; ARAÚJO, W.S. (eds.). *O uso ritual da ayahuasca*. 2nd ed. Campinas, Mercado de Letras. pp. 671-709.

ARAÚJO, M.C.R. (2005a). "Santo Daime: teoecologia e adaptação aos tempos modernos," Mestrado em Psicologia Social, Uerj.

ARAÚJO, M.C.R. (2005b). "Partilhando idéias: Santo Daime: teoecologia e adaptação aos tempos modernos," *Núcleo de Estudos Interdisciplinares sobre Psicoativos – Neip*, available at: www.neip.info/textos_colaboradores.htm, accessed September 2007.

BARBOSA, P.C.R. (2001). "Psiquiatria cultural do uso ritualizado de um alucinógeno no contexto urbano: uma investigação dos estados de consciência induzidos pela ingestão da ayahuasca no Santo Daime e União do Vegetal em moradores de São Paulo," Mestrado em Saúde Mental, Unicamp.

BARBOSA, P.C.R. (2006). "Uma avaliação neuropsicológica longitudinal de sujeitos que usam o alucinógeno amazônico-ameríndio ayahuasca em um contexto religioso-ritual urbano," Qualificação de Doutorado em Saúde Mental, Unicamp. (Research in progress)

BARBOSA, P.C.R.; DALGALARRONDO, P. (2003). "O uso ritual de um alucinógeno no contexto urbano: estados alterados de consciência e efeitos em curto prazo induzidos pela primeira experiência com a ayahuasca," *Jornal Brasileiro de Psiquiatria*, no. 52, vol. 3, pp. 181-190.

BARBOSA, P.C.R.; DALGALARRONDO, P.; GIGLIO, J. S. (2005). "Altered states of consciousness and short-term psychological after-effects induced by the first time ritual use of ayahuasca in an urban context in Brazil," *Journal of Psychoactive Drugs*, no. 37, vol. 2, pp. 193-201.

BRITO, G. de S. (2004). "Farmacologia humana da hoasca. Chá preparado de plantas alucinógenas usado em contexto ritual no Brasil," in: LABATE, B.C.; ARAÚJO, W.S. (eds.). *O uso ritual da ayahuasca*. 2nd ed. Campinas, Mercado de Letras, pp. 623-651.

CALLAWAY, J.C. (1994). "Another warning about harmala alkaloids and other MAO inhibitors," *MAPS Bulletin*, no. 4, vol. 4, available at: http://www.maps.org/news-letters/v04n4/04458mao.html, accessed September 2007.

CALLAWAY, J.C.; GROB, C.S. (1998). "Ayahuasca preparations and serotonin reuptake inhibitors: a potential combination for severe adverse interactions," *Journal of Psychoactive Drugs*, no. 30, vol. 4, pp. 367-369.

CALLAWAY, J.C.; AIRAKSINEN, M.M.; McKENNA, D.J.; BRITO, G.; GROB, C.S. (1994). "Platelet serotonin uptake sites increased in drinkers of ayahuasca," *Psychopharmacology*, no. 116, pp. 385-387.

CALLAWAY, J.C.; RAYMON, L.P.; HEARN, W.L.; McKENNA, D.J.; GROB, C.S.; BRITO, G.S. (1996). "Quantitation of N,N-dimethyltryptamine and harmala alkaloids in human plasma after oral dosing with ayahuasca," *Journal of Analytical Toxicology*, no. 20, pp. 492-97.

CALLAWAY, J.C.; McKENNA, D.J.; GROB, C.S.; BRITO, G.S.; RAYMON, L.P.; POLAND R.E.; ANDRADE E.N.; ANDRADE E.O.; MASH, D.C. (1999). "Pharmacokinetics of hoasca alkaloids in healthy humans," *Journal of Ethnopharmacology*, no. 65, pp. 243-256.

CALLAWAY, J.C.; BRITO, G.S.; NEVES, E.S. (2005). "Phytochemical analyses of Banisteriopsis caapi and Psychotria viridis," *Journal of Psychoactive Drugs*, no. 37, vol. 2, pp. 145-150.

CALLAWAY, J.C., GROB, C.S., McKENNA, D.J., NICHOLS, D.E. SHULGIN, A.; TUPPER, K.W. (2006). "A demand for clarity regarding a case report on the ingestion of 5-methoxy-N, N-dimethyltryptamine (5-MeO-DMT) in an ayahuasca preparation," *Journal of Analytical Toxicology*, no. 30, pp. 406-407.

CAMARGO, I.A. (2003). "El uso religioso del té ayahuasca y su relación con la psicosis: un estudio centralizado en la Unión del Vegetal y en el Santo Daime," *Maestria en Drogadependencias*, Universidad de Barcelona.

CARVALHO, J.E.; COSTA, M.; DIAS, P.C.; ANTÔNIO, M.A.; BARBOSA, F.U.; FOGLIO, M.A.; BRITO, A.R.M.S. (1995a). "Triagem fitoquímica e influência da hoasca sobre o consumo voluntário de etanol em camundongos," Trabalho apresentado na *I Conferência Internacional dos Estudos da Hoasca*. Rio de Janeiro.

CARVALHO, J.E., COSTA, M., DIAS, P.C., ANTÔNIO, M.A.; BRITO, A.R.M.S. (1995b). "Efeitos farmacológicos do decocto (Hoasca) de Banisteriopsis caapi e Psychotria viridis em camundongos," Trabalho apresentado na I *Conferência Internacional dos Estudos da Hoasca*. Rio de Janeiro.

CARVALHO, T.B. (2005). "Em busca do encontro: a demanda numinosa no contexto religioso da União do Vegetal," Mestrado em Psicologia, PUC-RJ.

CAZENAVE, S.O.S. (2000). "Banisteriopsis caapi e ação alucinógena," *Revista de Psiquiatria clínica*, no. 27, vol. 1.

CENTRO ESPÍRITA BENEFICENTE UNIÃO DO VEGETAL (CEBUDV) (2003). "Pesquisa com adolescentes exige mobilização," *Alto Falante* (edição eletrônica), CEBUDV, nov. 2003, p. 1.

CENTRO ESPÍRITA BENEFICENTE UNIÃO DO VEGETAL (CEBUDV) (2004a). "Pesquisa com adolescentes – arrecadação é vitoriosa," *Alto Falante* (edição eletrônica), CEBUDV, fev. 2004, p. 1.

CENTRO ESPÍRITA BENEFICENTE UNIÃO DO VEGETAL (CEBUDV) (2004b). Pesquisa em desenvolvimento por duas universidades. *Alto Falante* (edição eletrônica), CEBUDV, fev. 2004, p. 2.

CHAVES, L. R. (2003). "A mulher urbana no Santo Daime: entre o modelo arcaico e o moderno de feminino," Mestrado em Psicossociologia de Comunidade e Ecologia Social, UFRJ.

CONSELHO NACIONAL ANTIDROGAS (2003). Resolução no. 26 de 31 de dezembro de 2002, *Diário Oficial da União* 01/01/2003, Seção 1, Brasília.

CONSELHO NACIONAL ANTIDROGAS (2004). Resolução no. 4 de 4 de novembro de 2004, Diário Oficial da União 08/11/2004, Seção 1, Brasília.

CONSELHO NACIONAL ANTIDROGAS (2006). Relatório Final do Grupo Multidisciplinar de Trabalho – GMT Ayahuasca, Brasília.

COSTA, M.C.M.; FIGUEIREDO, M.C.; CAZENAVE, S.O.S. (2005). "Ayahuasca: uma abordagem toxicológica do uso ritualístico," *Revista de Psiquiatria clínica*, no. 32, vol. 6.

COUGAR, M. (2005). "An investigation of personal transformations and psychoactive plant use in syncretic ritual ceremonies in a Brazilian church," PhD Thesis in Transpersonal Psychology, Institute of Transpersonal Psychology.

DA SILVEIRA, D.X.; GROB, C.S.; DOBKIN DE RIOS, M.; LOPEZ, E.; ALONSO, L.K.; TACLA, C.; DOERING-SILVEIRA, E. (2005). "Ayahuasca in adolescence: a preliminary psychiatric assessment," *Journal of Psychoactive Drugs*, no. 37, vol. 2, pp. 129-134.

DA SILVEIRA, D.X. (2007). "O uso de ayahuasca em contexto ritual religioso," *VIII Fórum Internacional em Saúde – "As drogas lícitas e ilícitas na Amazônia legal: uso, abuso e recuperação,"* Rio Branco.

DÁVILA, M. (2007). "Voluntário número 13," *Revista Piauí,* April, available at: http://www.revistapiaui.com.br/2007/abr/fisico.htm, accessed September 2007.

DIAS, M. (2007). "A pesquisa tem `mironga': notas etnográficas sobre o fazer etnográfico," in: Bonetti, A.; Fleischer, S. (eds.). *Entre saias justas e jogos de cintura.* Florianópolis, Editora Mulheres/Edunisc. pp. 73-92.

DOBKIN DE RIOS, M.; GROB, C.S.; LOPEZ, E.; DA SILVIERA, D.X.; ALONSO, L.K.; DOERING-SILVEIRA, E. (2005). "Ayahuasca in adolescence: qualitative results," *Journal of Psychoactive Drugs*, no. 37, vol. 2, p. 135-140.

DOERING-SILVEIRA, E. (2003). "Avaliação neuropsicológica de adolescentes que consomem chá de ayahuasca em contexto ritual religioso," Mestrado em Psiquiatria e Psicologia Médica, Unifesp/EPM.

DOERING-SILVEIRA, E.; LOPEZ, E.; GROB, C.S.; DOBKIN DE R. M.; ALON-

SO, L.K.; TACLA, C.; SHIRAKAWA, I.; BERTOLUCCI, P.H.; DA SILVEIRA, D.X. (2005a). "Ayahuasca in adolescence: a neuropsychological assessment," *Journal of Psychoactive Drugs*, no. 37, vol. 2, pp. 123-128.

DOERING-SILVEIRA, E.; GROB, C.S.; DOBKIN DE RIOS, M.; LOPEZ, E.; ALONSO, L.K.; TACLA, C.; DA SILVEIRA, D.X. (2005b). "Report on psychoactive drug use among adolescents using ayahuasca within a religious context," *Journal of Psychoactive Drugs*, no. 37, vol. 2, pp. 141-144.

GOULART, S.L. (2004). "Contrastes e continuidades em uma tradição amazônica: as religiões da ayahuasca," Doutorado em Antropologia, Unicamp.

GRINSPOON, L.; BAKALAR, J.B. (1981). *Psychedelic drugs reconsidered*. New York, Basic Books.

GROB, C.S. (2002). *Hallucinogens: a reader*. New York, Tarcher/Putnam.

GROB, C.S.; McKENNA, D.J.; CALLAWAY, J.C.; BRITO, G.S.; NEVES, E.S.; OBERLENDER, G.; SAIDE, O.L.; LABIGALINI JR., E.; TACLA, C.; MIRANDA, C.T.; STRASSMAN, R.J.; BOONE, K.B. (1996a). "Human psychopharmacology of hoasca, a plant hallucinogen used in ritual context in Brazil," *Journal of Nervous & Mental Disease*, no.. 184, vol. 2, pp. 86-94.

GROB, C.S.; McKENNA, D.J.; CALLAWAY, J.C.; BRITO, G.S.; NEVES, E.S.; OBERLENDER, G.; SAIDE, O.L.; LABIGALINI JR., E.; TACLA, C.; MIRANDA, C.T.; STRASSMAN, R.J.; BOONE, K.B. (1996b). "Farmacologia humana da hoasca, planta alucinógena usada em contexto ritual no Brasil: I. Efeitos psicológicos," *Informação Psiquiátrica*, no. 15, vol. 2, pp. 39-45. (tradução do artigo Grob et. al. 1996 para o português)

GROB, C.S.; McKENNA, D.J.; CALLAWAY, J.C.; BRITO, G.S.; ANDRADE, E.O.; OBERLENDER, G.; A, O.L.; LABIGALINI JR., E.; TACLA, C.; MIRANDA, C.T.; STRASSMAN, R.J.; BOONE, K.B.; NEVES, E.S. (2004). "Farmacologia humana da hoasca, planta alucinógena usada em contexto ritual no Brasil: efeitos psicológicos," in: LABATE, B.C.; ARAÚJO, W.S. (eds.). *O uso ritual da ayahuasca*. 2nd ed., Campinas, Mercado de Letras, pp. 653-669. (tradução do artigo Grob et. al. 1996 para o português)

GROF, S. (2001). *LSD psychotherapy,* Sarasota, Multidisciplinary Association for Psychedelic Studies (MAPS).

JANSEN, K. (2004). *Ketamine: dreams and realities,* Sarasota, Multidisciplinary Association for Psychedelic Studies (MAPS).

LABATE, B.C. (2004). *A reinvenção do uso da ayahuasca nos centros urbanos.* Campinas, Mercado de Letras.

LABATE, B.C. (2005). "Dimensões legais, éticas e políticas da expansão do consumo da ayahuasca," in: LABATE, B.C.; GOULART, S.L. (eds.). *O uso ritual das plantas de poder.* Campinas, Mercado de Letras, pp. 397-457.

LABATE, B.C.; Araújo, W.S. (eds.) (2004). *O uso ritual da ayahuasca.* Campinas, Mercado de Letras, 2a ed.

LABATE, B.C.; GOULART, S.L. (eds.) (2005). *O uso ritual das plantas de poder.* Campinas, Mercado de Letras.

LABATE, B.C.; GOULART, S.L.; CARNEIRO, H.S. (2005). "Introdução," in: LABATE, B.C.; GOULART, S.L. (eds.). *O uso ritual das plantas de poder.* Campinas, Mercado de Letras, pp. 29-55.

LABIGALINI JR., E. (1998). "O uso de ayahuasca em um contexto religioso por ex-dependentes de álcool: um estudo qualitativo," Mestrado em Saúde Mental, Unifesp/EPM.

LABIGALINI JR., E.; DUNN, J. (1995). "The Union of the Vegetable: the ritualized use of hoasca tea," *Psychiatric Bulletin,* no. 19, pp. 313-314.

LABIGALINI JR., E.; RODRIGUES, L.R (1997). "O uso 'terapêutico' de Cannabis por dependentes de crack no Brasil." *Psychiatric On-Line Brazil,* no. 2, vol. 2, available at: http://www.priory.com/psych/eliseu.htm, acessado em setembro de 2007.

LATOUR, B. (1994). *Jamais fomos modernos: ensaio de antropologia simétrica.* São Paulo, Editora 34

LIMA, F.A. de S. (1996-1997). "The ritual use of hoasca: comments and advice," *MAPS Bulletin*, vol. 7, no. 1, pp. 25-26, available at: http://www.maps.org/news-letters/v07n1/07125udv.html, accessed September 2007.

LIMA, F.A. de S.; NAVES, M.B.; MOTTA, J.M.C.; DI MIGUELI, J.C.V.; BRITO, G.S. et al. (1998). "Sistema de notificação e monitoramento psiquiátrico em instituição de usuários do chá hoasca – União da Vegetal," *XVI Congresso Brasileiro de Psiquiatria*, São Paulo.

LIMA, F.A. DE S.; NAVES, M.B.; MOTTA, J.M.C.; DI MIGUELI, J.C.V.; BRITO, G.S. et al. (2002). "Sistema de monitoramento psiquiátrico de usuários do chá hoasca," *Revista Brasileira de Psiquiatria*, no. 24 (supl. 2).

MARRAS, S. (2002). "Ratos e homens – e o efeito placebo: um reencontro da cultura no caminho da natureza," *Campos*, no. 2, pp. 117-133.

MARTINS-VELLOSO, S. (2003). "Imaginaire collectif, productions oniriques et modalités de consommation de l'ayahuasca en forêt amazonienne," *Mémoire de fin de Capacité en Addictologie Clinique*, Université Paris, no. 5.

MASTERS, R.; HOUSTON, J. (2000). *The varieties of psychedelic experience: the classic guide to the effects of LSD on the human psyche*. Rochester, VT, Park Street Press.

McKENNA, D.J. (2004). "Clinical investigations of the therapeutic potential of ayahuasca: rationale and regulatory challenges," *Pharmacology & Therapeutics*, no. 102, pp. 111-129.

McKENNA, D.J.; CALLAWAY, J.C.; GROB, C.S. (1998). "The scientific investigation of Ayahuasca: a review of past and current research," *The Heffter Review of Psychedelic Research*, no. 1, pp. 65-77.

MENZE, P. (2004). "Ayahuasca in rituele en therapeutische context," Doctoraalscriptie Psychologie, Rijksuniversiteit Utrecht.

MEYER, M. (2006). "Religious freedom and United States drug laws: notes on the UDV-USA legal case," *Núcleo de Estudos Interdisciplinares sobre Psicoativos – Neip*, available at: http://www.neip.info/ downloads/Matthew%20UDV-USA%20case.pdf, accessed September 2007.

MIRANDA, C.T.; LABIGALINI JR., E.; TACLA, C. (1995). "Alternative religion and outcome of alcohol dependence in Brazil," *Addiction,* no. 90, vol. 6, p. 847.

MORAIS, A.F. (2005). "O ethos e o futuro na Vila Céu do Mapiá, Amazonas," Doutorado em Psicologia Social, USP.

NOVAES, C. (2006). "L'expérience de l'ayahuasca et ses `états modifiés de la conscience'. Une étude transculturelle des récits des usagers urbains de l'ayahuasca. Une lecture a travers le concept de l'inconscient selon Gilles Deleuze," Master recherche Développement, Psychopathologie et Psychanalyse, Clinique Transculturelle, Option Clinique Transculturelle, Université Paris, no. 13.

OTT, J. (1994). *Ayahuasca analogues: pangaean entheogens.* Kennewick, Natural Books Co.

OTT, J. (1998). *Pharmacophilia o los paraísos naturales.* Barcelona, Phantastica.

PELÁEZ, M.C. (1994). "No mundo se cura tudo: interpretações sobre a "cura espiritual" no Santo Daime," Mestrado em Antropologia Social, UFSC.

PIGNARRE, P. (1999). O que é o medicamento? Um objeto estranho entre ciência, mercado e sociedade. São Paulo, Editora 34.

PRASCINA, A. (1997). "Origine e sviluppo del culto amazzonico del Santo Daime," Tesi di Laurea in Psicologia, Università degli Studi di Padova.

QUINLAN, M. (2001). "Healing from the Gods: ayahuasca and the curing of disease states," PhD Thesis in Integral Studies, California Institute of Integral Studies.

RIBA, J. (2003). "Human pharmacology of Ayahuasca," Tesi Doctoral Dissertation, Universitat Autonoma de Barcelona.

RIBA, J.; BARBANOJ, M.J. (2005). "Bringing ayahuasca to the clinical research laboratory," *Journal of Psychoactive Drugs,* no. 37, vol. 2, pp. 219-230.

RIBA, J.; RODRIGUES-FORNELLS, A.; STRASSMAN, R.J.; BARBANOJ, M.J.

(2001a). "Psychometric assessment of the hallucinogen rating scale," *Drug and Alcohol Dependence*, no. 62, pp. 215-223.

RIBA, J.; RODRIGUES-FORNELLS, A.; URBANO, G.; MORTE, A.; ANTONI-JOAN, R.; MONTEIRO, M.; CALLAWAY, J.C.; BARBANOJ, M.J. (2001b). "Subjective effects and tolerability of the South American psychoactive beverage Ayahuasca in healthy volunteers," *Psychopharmacology* (Berl), no. 154, pp. 85-95.

RIBA, J.; RODRIGUES-FORNELLS, A.; BARBANOJ, M.J. (2002a). "Effects of ayahuasca sensory and sensorimotor gating in humans as measured by P50 suppression and prepulse inhibition of the startle reflex, respectively," *Psychopharmacology* (Berl), no. 165, pp. 18-28.

RIBA, J.; ANDERER, P.; MORTE, A.; URBANO, G.; JANE, F.; SALETU, B.; BARBANOJ, M.J. (2002b). "Topographic pharmaco-EEG mapping of the effects of the South American beverage ayahuasca in healthy volunteers," *British Journal of Clinical Pharmacology*, no. 5, pp. 3613-628.

RIBA, J.; VALLE, M.; URBANO, G.; YRITIA, M.; MORTE, A.; BARBANOJ, M.J. (2003). "Human pharmacology of ayahuasca: subjective and cardiovascular effects, monoamine metabolite excretion, and pharmacokinetics," *Journal of Pharmacology and Experimental Therapeutics*, no. 306, pp. 73-83.

RIBA, J.; ANDERER, P.; JANÉ, F.; SALETU, B; BARBANOJ, M.J. (2004). "Effects of the South American psychoactive beverage ayahuasca on regional brain electrical activity in humans: a functional neuroimaging study using low-resolution electromagnetic tomography," *Neuropsychobiology*, no. 50, pp. 89-101.

RIBA, J.; ROMERO, S.; GRASA, E.; MENA, E.; CARRIÓ, I.; BARBANOJ, M.J. (2006). "Increased frontal and paralimbic activation following ayahuasca, the pan-Amazonian inebriant," *Psychopharmacology*, no. 186, pp. 93-98.

RODRIGUES, L.R. (2006). "Uso religioso e ritual de substâncias psicoativas," in: SILVEIRA, D.X. DA; MOREIRA, F.G. (eds.). *Panorama atual de drogas e dependências*. São Paulo, Atheneu. pp. 435-439.

ROSE, I.S. de (2005). "Espiritualidade, terapia e cura: um estudo sobre a expressão da experiência no Santo Daime," Mestrado em Antropologia Social, UFSC.

SANTOS, R.G. dos (2006). "Efeitos da ingestão de ayahuasca em estados psicométricos relacionados ao pânico, ansiedade e depressão em membros do culto Santo Daime," Mestrado em Psicologia, Processos Comportamentais, UNB.

SANTOS, R.G. dos (2007a). "Ayahuasca: neuroquímica e farmacologia," *Smad – Revista Eletrônica Saúde Mental Álcool e Drogas,* no. 3, vol. 1, available at http://www2.eerp.usp.br/resmad/artigo_titulo.asp?rmr=104, accessed September 2007.

SANTOS, R. G. dos (2007b). "Estudo sobre o mecanismo de ação da Ayahuasca em humanos: pré-tratamento com ketanserina," Doutorado em Farmacologia, Universitat Autònoma de Barcelona. (Research in progress)

SANTOS, R. G. DOS; LANDEIRA-FERNANDEZ, J.; STRASSMAN, R.J.; MOTTA, V.; CRUZ, A.P.M. (2007). "Effects of Ayahuasca on psychometric measures of anxiety, panic-like and hopelessness in Santo Daime members," *Journal of Ethnopharmacology,* no. 112, vol. 3, pp. 50 7-513.

SCHULTES, R.E. (1986). "El desarrollo histórico de la identificación de las malpigiáceas empleadas como alucinógenos," *América Indígena,* no. 46, vol. 1, pp. 319-330.

SHANON, B. (2002). *The antipodes of the mind: charting the phenomenology of the ayahuasca experience.* New York, Oxford University Press.

SHANON, B. (2003). "Os conteúdos das visões da ayahuasca," *Mana,* no. 9, vol 2, pp. 109-152.

SHANON, B. (2004). "A ayahuasca e o estudo da mente," in: LABATE, B.C.; ARAÚJO, W.S. (eds.). *O uso ritual da ayahuasca.* 2nd ed. Campinas, Mercado de Letras. pp. 681-709.

SKLEROV, J.; LEVINE, B.; MOORE, K.A.; KING, T.; FOWLER, D. (2005). "A fatal intoxication following the ingestion of 5-Methoxy-N, N-Dimethyltryptamine in an Ayahuasca preparation," *Journal of Analytical Toxicology*, no. 29, pp. 838-841.

STRASSMAN, R.J. (1984). "Adverse reactions to psychedelic drugs: a review of the literature," *Journal of Nervous and Mental Disease*, no. 172, vol. 10, pp. 577-595.

STRASSMAN, R.J. (2001). DMT: the spirit molecule. Rochester, VT, Park Street Press.

STRASSMAN, R.J.; QUALLS, C.R. (1994). "Dose-response study of N, N-dimethyltryptamine in humans: I. Neuroendocrine, autonomic, and cardiovascular effects," *Archives of General Psychiatry*, vol. 51, pp. 85-97.

STRASSMAN, R.J.; QUALLS, C.R.; UHLENHUTH, E.H.; KELLNER, R. (1994). "Dose-response study of N, N-dimethyltryptamine in humans: II. Subjective effects and preliminary results of a new rating scale," *Archives of General Psychiatry*, no. 51, pp. 98-108.

STRASSMAN, R.J.; QUALLS, C.R.; BERG, L.M. (1996). "Differential tolerance to biological and subjective effects of four closely spaced doses of N, N-dimethyltryptamine in humans," *Biological Psychiatry*, no. 39, vol. 9, pp. 784-795.

SULLA, J. (2005). "The system of healing used in the Santo Daime community Céu do Mapiá," Master of Arts in Psychology, Saybrook Graduate School and Research Center.

UMIYAC (Unión de Médicos Indígenas Yageceros de Colombia) (1999). *Encuentro de Taitas en la Amazonía colombiana. Ceremonias y reflexiones*. Caquetá, Colombia, June 1-8.

VILLAESCUSA, M. (2002). "An exploration of psychotherapeutic aspects of Santo Daime ceremonies in the UK," MSc in Humanistic Integrative Psychotherapy, Middlesex University.

VILLAESCUSA, M. (2003). "Aspetos psicoterapeuticos de las ceremonias del Santo Daime en el Reino Unido," in: Fericgla, J.M. (ed.). *BI - Boletín de la*

Sociedad de Etnopsicología Aplicada. Monográfico Ayahuasca. Barcelona, Sd'EA. pp. 40-50.

YRITIA, M.; RIBA, J.; ORTUNO, J.; RAMIREZ, A.; CASTILLO, A.; ALFARO, Y.; DE LA TORRE, R.; BARBANOJ, M.J. (2002). "Determination of N, N-dimethyltryptamine and â-carboline alkaloids in human plasma following oral administration of Ayahuasca," *Journal of Chromatography B*, no. 779, pp. 271-281.

ZINBERG, N. (1984). *Drug, set and setting.* New Haven, Yale University Press.

Mestre Moacir Biondo during a regular session of
the Centro Espírita Beneficente União do Vegetal (UDV),
Núcleo Águas Claras, Manaus (Amazonas state)
Crédits: Photo by Henrique Biondo

Núcleo Caupuri, Centro Espírita Beneficente União do Vegetal (UDV), Manaus (Amazonas state).
Credits: Photo by Beatriz C. Labate

Dancing Work – Festival of the Holy Kings – Cefluris/Santo Daime Church,
Céu do Mapiá, Amazônia – 2007
Credits: Photo by Andréa D'Amato

Opening the work in the terreiro, feast commemorating the day of Dom Simeão, held November 23, 2005 at the Centro Espírita Obras de Caridade Príncipe Espadarte (Barquinha church) in Rio Branco (Acre state). Credits: Photo by Yledo Junior

Padrinho Alfredo Gregório de Melo, world leader of Cefluris/Santo Daime
Credits: Evelyn Ruman

Chapter 3

Bibliography of the Ayahuasca Religions[71]

This chapter's objective is to present a list of bibliographical references on the so-called "Brazilian ayahuasca religions"—Santo Daime (in its Alto Santo and Cefluris denominations), the União do Vegetal (UDV), and the Barquinha—which is intended to be the most exhaustive possible, and which aims to provide an overview of the state of the art of this literature and a useful guide for researchers in the area.

The references are organized by language: Danish, Dutch, English, French, German, Italian, Japanese, Norwegian, Portuguese, and Spanish. The criteria for selecting the Portuguese language references called for the inclusion of post-graduate work (master's and PhD) either finished or in progress, whether published or not in book form, and scientific articles published in books, journals, or online. We also chose to list writings of a non-academic nature produced by practitioners of the ayahuasca religions, and to include works in the areas of pharmacology, psychology, and biomedicine that were conducted in the context of these religions, or which consider their more symbolic aspects. In this case, we have indicated, where possible, the locale of fieldwork

---

71. Govert Derix assisted the preparation of the text. It is an expanded version of the "Bibliography of the Brazilian Ayahuasca religions" written by Beatriz Caiuby Labate, Rafael Guimarães dos Santos, Isabel Santana de Rose and Govert Derix and published in June, 2007 on the website of the Multidisciplinary Association for Psychedelic Studies (MAPS), available at http://www.maps.org/Ayahuasca/ Ayahuascabibliography.pdf (For a brief commentary on the compilation of the original bibliography, see Labate et al., 2007). This version, concluded in November, 2007, added new references and a series of links pertaining to the articles cited in the list which lead either to the texts themselves or to additional information about them. All the links given were accessed in September, 2007.

in brackets alongside the citation. In languages other than Portuguese, given the scarcity of writings on the topic, the criteria for including references were a bit broader, including bachelor's theses, articles published in popular magazines, and books, theses, and articles that discuss the Brazilian ayahuasca religions only indirectly.[72] Legal documents and news articles about legal cases involving the União do Vegetal and Santo Daime in Brazil, the United States, and Europe were excluded.

Danish

DAMM, I. (1999). "Til te med regnskovens guder trylledrikk," *Illustrert videnskab*, no. 8, pp. 70-75.

Dutch

BOGERS, H. (1994). "Ayahuasca: la purga: de klimplant van de kleine dood," in: PLOMP, H.; BOGERS, H.; SNELDERS, S. *De psychedelische (r)evolutie*. Amsterdam, Editora Bres, pp. 215-253.

BOGERS, H. (1995). "De Santo Daime leer. Ayahuascagebruik in een religieuze setting", *PAN Forum,* no. 1, pp. 2-10.

BOGERS, H. (1996). "Gesprek met de spiritueel leider van Santo Daime: Padrinho Alfredo Gregório Mota de Melo", *PAN Forum,* no. 4, pp. 22-27.

BOGERS, H. (1997-1998). "De planten der Goden in de 21ste eeuw", *Bres,* no. 187, pp. 55-64.

BOGERS, H. (2001). "Duizend jaar in één nacht", in: PAN (Psycho Actief Netwerk) (org). *Psychedelische perspectieven. De renaissance van de trip in de 21ste eeuw.* Amsterdam, Editora Bres, pp. 81-91.

BOGERS, H. (2001). "Eigentijdse inquisitie in Nederland. Santo Daime: de vervolging van een entheogene religie", *Bres,* no. 206, pp. 33-43.

BOGERS, H. (2002). "Santo Daime legal!", *Bres,* no. 213, pp. 14-24.

---

72. We have opted not to translate the academic degrees associated with post-graduate works, given the difficulty of establishing universal parallels among the various academic systems, which impeded a "cultural translation" of them all (for example, are the French "master recherché," the Brazilian "dissertação de mestrado, and the Italian "tesi di laurea" equivalent?). These terms were left in the original form, which should facilitate the localization of references by the reader.

DERIX, G. (2003). "Het begin van een beweging. Beschouwing naar aanleiding van de eerste Europese conferentie over Ayahuasca", *Bres*, no. 219, pp. 86-98.

DERIX, G. (2004a). "Ayahuasca. Een kritiek van de psychedelische rede. Avontuur in het Amazonegebied," Amsterdam/Antwerp, De Arbeiderspers, available at: http://alto-das-estrelas.blogspot.com/2006/12/livro-em-alemo-e-holands-sobre.html and http://alto-das-estrelas.blogspot.com/2006/12/summary-in-english-of-govert-derixs.html.

DERIX, G. (2004b). "Waarom Ayahuasca drinken? Wat de ayahuasqueiros bezielt," *Bres*, no. 228, pp. 74-79.

LEMMENS, P. (2006). Het verstand loslaten. Recensie van: Ayahuasca. Een kritiek van de psychedelische rede. *Filosofie & Praktijk,* no. 4, pp. 56-59.

MENZE, P. (2004). "Ayahuasca in rituele en therapeutische context," Doctoraalscriptie Psychologie, Rijksuniversiteit Utrecht.

NAPEL, T. (2002). "De Santo Daime kerk en het recht op anders-zijn," *Nederlands tijdschrift voor de mensenrechten,* no. 27, vol. 6, pp. 723-727.

NOORMAN, L. (1999). "Telma, daimista," in: Noorman, L. *Ankerplaats Amazona,* Amsterdam, Nijgh en van Ditmar, pp. 189-221.

REUS, A. de (2002). "De eredienst van Santo Daime. Drugs als weg naar God." *Onkruid,* nov-dec, pp. 8-17.

VERBURG, P.G.F. (1994). "Santo Daime: sacrament uit het regenwoud." *Bres,* no. 165, pp. 39-48.

WUYTS, J. (2007). "Urbaan sjamanisme; een case-study. De ontwikkeling van de Santo Daime beweging, van Amazonewoud tot stedelijke setting en verder" (Provisional title), Master scriptie Culturele Antropologie en Ontwikkelingssociologie, Universiteit Leiden. (Project in development)

English

ADELAARS, A. (1997-1998). "Psychedelic rituals in the Netherlands," *Yearbook for Ethnomedicine and the Study of Consciousness,* no. 6-7, pp. 355-340, available at: http://www.xs4all.nl/~nota/.

ALVERGA, A.P. de (1999). *Forest of visions. Ayahuasca, Amazonian spirituality, and the Santo Daime tradition*, Rochester, VT, Park Street Press.

ALVERGA, A.P. de (2000). "The book of visions: journey to Santo Daime (excerpt)," in: Luna, L.E.; White, S. (eds.). *Ayahuasca reader. Encounters with the Amazon's sacred vine*. Santa Fe, NM, Synergetic Press, pp. 145-153.

ANDERSON, B. (2007). "Enchantment and environment: environmental values in the Centro Espírita Beneficente União do Vegetal," Undergraduate independent study in Latin American and Latino Studies, University of Pennsylvania.

ANDRADE, R.G.; CANABRAVA, V. (2002). "Communities of Santo Daime in Brazil," XXIII Conference and General Assembly IAMCR/AIECS/AIERI. Barcelona, available at: http://www.portalcomunicacion.com/bcn2002/n_eng/programme/prog_ind/papers/0_arribats_peremail/abans_07_2002/pdf/andrade.pdf.

BALZER, C. (2005). "Ayahuasca rituals in Germany: the first steps of the Brazilian Santo Daime religion in Europe," *Curare: Zeitschrift für ethnomedizin und transkulturelle psychiatrie*, no. 28, vol. 1, pp. 53-66.

BARBOSA, P.C.R.; DALGALARRONDO, P.; GIGLIO, J.S. (2005). "Altered states of consciousness and short-term psychological after-effects induced by the first time ritual use of Ayahuasca in urban context in brazil," *Journal of Psychoactive Drugs*, no. 37, vol. 2, pp. 193-201. [Results of research with first-time users of ayahuasca nineteen people in Santo Daime in São Paulo and nine people in the UDV in Campinas and São Paulo) conducted in 1997 and 2000.]

BIANCHI, A. (2004). Review of: *O uso ritual da Ayahuasca. Organizzazione Interdiciplinare Sviluppo e Sallute* – ORISS, 2004, available at: http://www.oriss.org/articoli/a_Ayahuasca_en.html e http://www.Neip.info/downloads/l_bia1_antonio_bianchi.htm.

BOIRE, R.G. (2006). "RGB on UDV vs USA: Notes on the hoasca supreme court decision," *Multidisciplinary Associataion of Psychedelic Studies Bulletin*, no. 16, vol. 1, pp. 19-21, available at: http://www.maps.org/news-letters/v16n1-html/index.html.

CALLAWAY, J.C.; AIRAKSINEN, M.M.; MCKENNA, D.J.; BRITO, G.; GROB, C.S. (1994). "Platelet serotonin uptake sites increased in drinkers of Ayahuasca," *Psychopharmacology*, no. 116, pp. 385-387. [Results of the Hoasca Project, research conducted with long-term users of ayahuasca fifteen members of the UDV with at least ten years' experience) in Manaus, Amazonas state, in 1993.]

CALLAWAY, J.C.; RAYMON, L.P.; HEARN, W.L.; McKENNA, D.J.; GROB, C.S.; BRITO, G.S. (1996). "Quantitation of N,N-dimethyltryptamine and harmala alkaloids in human plasma after oral dosing with Ayahuasca," *Journal of Analytical Toxicology*, no. 20, pp. 492-97. [Results of the Hoasca Project.]

CALLAWAY, J.C.; McKENNA, D.J.; GROB, C.S.; BRITO, G. S.; RAYMON, L.P.; POLAND, R.E.; ANDRADE, E.N.; ANDRADE, E.O.; MASH, D.C. (1999). "Pharmacokinetics of Hoasca alkaloids in healthy humans," *Journal of Ethnopharmacology*, no. 65, pp. 243-256. [Results of the Hoasca Project], available at: http://www.udv.org.br/portugues/downloads/04.pdf.

CAMURÇA, M. A. (2005). Review of: A reinvenção do uso da Ayahuasca nos centros urbanos. Multidisciplinary Association for Psychedelic Studies – MAPS, available at: http://www.maps.org/reviews/reinvention.html and http://www.Neip.info/downloads/resenha %20Marcelo%20Camura%20A%20reinveno%20ingles.pdf.

CARNEIRO, H. (2002-2003). Review of: O uso ritual da Ayahuasca. Multidisciplinary Association for Psychedelic Studies– MAPS, available at: http://www.maps.org/reviews/rua.html and http://www.neip.info/ downloads/l_bia1_Henrique_ing.htm.

CARVALHO, J.J. De (2003). "The mystic of the marginal spirits," in: Torres, Y.G.; Pye, M. (2003). Religion and Society: Proceedings of the 17th Quinquennial Congress of the International Association for the History of Religions (IAHR). Cambridge, *Roots and Branches*, pp. 71-108. [English translation of the original Spanish-language article "El misticismo de los espíritus marginales."]

CHAUMEIL, J.P. (1992). "Varieties of Amazonian Shamanism," *Diogenes*, no. 158, pp. 101-113.

COUGAR, M. (2005). "An investigation of personal transformations and psychoactive plant use in syncretic ritual ceremonies in a Brazilian church," PhD Thesis in Transpersonal Psychology, Institute of Transpersonal Psychology. [Results of research with fifty-two North Americans and Europeans who visited Brazil, principally Céu do Mapiá / Santo Daime-Cefluris, during the December festivals, from 1999 to 2002.] Available at: http://www.neip. info/downloads/Santo%20Daime%20Paper.pdf.

DA SILVEIRA, D.X.; GROB, C.S.; DOBKIN DE RIOS, M.; LOPEZ, E.; ALONSO, L.K.; TACLA, C.; DOERING-SILVEIRA, E. (2001). "Ayahuasca in adolescence: a preliminary psychiatric assessment," *Journal of Psychoactive Drugs*, vol. 37, no. 2, pp. 129-134, [Results of research with forty adolescents from the UDV in three Brazilian cities (São Paulo, Campinas, and Brasília)] available at: http://www.Neip.info/downloads/adoles_udv/arquivos/da%20 Silveira_ps%20PM372.pdf.

DAWSON, A. (2007). "The Ayahuasca religions of Brazil," in: Dawson, A. *New Era – New Religions: religious transformation in contemporary Brazil*. Aldershot, Ashgate, pp. 67-98, (in press), available at: http://alto-das-estrelas.blogspot.com/2007/05/livro-sobre-movimento-nova-era-no.html.

DAWSON, A. (2007). "The Emergence, spread and diversification of Santo Daime (Cefluris) in Brazil," Research sponsored by The British Academy and The Leverhulme Trust, conducted at the Universidade of Lancaster. (Project in development) Available at: http://alto-das-estrelas.blogspot.com/2007/05/pesquisa-em-andamento-sobre-o-santo.html.

DOBKIN DE RIOS, M. (1996). "Commentary on 'human pharmacology of Hoasca': a medical anthropological perspective," *The Journal of Nervous and Mental Disease*, no. 184, vol. 2, pp. 95-98. [Comments on the results of the Hoasca Project]

DOBKIN DE RIOS, M.; GROB, C.S. (2005). "Editors' introduction: Ayahuasca use in cross-cultural perspective," *Journal of Psychoactive Drugs,* no. 37, no. 2, pp. 119-121.

DOBKIN DE RIOS, M.; GROB, C.S.; LOPEZ, E.; DA SILVIERA, D.X.; ALONSO, L.K.; DOERING-SILVEIRA, E. (2005). "Ayahuasca in adolescence: qualitative results," *Journal of Psychoactive Drugs,* no. 37, vol. 2, pp. 135-140. [Results of research with twenty-eight adolescents from the UDV in three Brazilian cities (São Paulo, Campinas e Brasília), 2001], available at: http://www.Neip.info/downloads/adoles_udv/arquivos/de%20Rios%20PM372.pdf.

DOBKIN DE RIOS, M.; GROB, C.S. (2005). "Interview with Jeffrey Bronfman, representative mestre for the União do Vegetal church in the United States," *Journal of Psychoactive Drugs,* no. 37, vol. 2, pp. 189-191.

DOERING-SILVEIRA, E.; LOPEZ, E.; GROB, C.S.; DOBKIN DE RIOS, M.; ALONSO, L.K.; TACLA, C.; SHIRAKAWA, I.; BERTOLUCCI, P.H.; DA SILVEIRA, D.X. (2005). "Ayahuasca in adolescence: a neuropsychological assessment," *Journal of Psychoactive Drugs,* no. 37, vol. 2, pp. 123-128. [Results of research with forty adolescent from the UDV in three Brazilian cities (São Paulo, Campinas, and Brasília), 2001], available at: http://www.Neip.info/downloads/adoles_udv/arquivos/da%20Silveira_ne%20PM372.pdf.

DOERING-SILVEIRA, E.; GROB, C.S.; DOBKIN DE RIOS, M.; LOPEZ, E.; ALONSO, L.K.; TACLA, C.; DA SILVEIRA, D.X. (2005). "Report on psychoactive drug use among adolescents using Ayahuasca within a religious context," *Journal of Psychoactive Drugs,* no. 37, vol. 2, pp. 141-144, [Results of research with forty-one adolescentes da União do Vegetal de três cidades do Brasil (São Paulo, Campinas e Brasília), 2001], available at: http://www.Neip.info/downloads/adoles_udv/arquivos/Doering%20PM372.pdf.

FRENOPOULO, C. (2004). "The mechanics of religious synthesis in the Barquinha religion," *Rever: Revista de Estudos da Religião,* no. 4, vol. 1, pp. 19-40, available at: http://www.Neip.info/downloads/t_christ1.pdf.

FRENOPOULO, C. (2005a). "Charity and spirits in the Amazonian navy: the

Barquinha mission of the Brazilian Amazon," Master of Arts in Anthropology, University of Regina.

FRENOPOULO, C. (2005b). Healing performances in the Barquinha religion: the tropes of exorcisms, passes and pontos riscados. Society for the Anthropology of Religion Meeting, Vancouver.

FRENOPOULO, C. (2005c). Review of: O uso ritual da Ayahuasca, *Journal of Psychoactive Drugs,* no. 37, vol. 2, pp. 237-239, available at: http://www.Neip.info/downloads/resenha %20Christian%20Frenolupo%20Journal%20of%20Psychoactive%20Drugs.pdf.

GOW, P. (2004). Review of: O uso ritual da Ayahuasca. *American Anthropologist,* no. 106, vol. 3, pp. 624, available at: http://www.neip.info/downloads/l_bia1_peter.htm.

GROB, C.S.; McKENNA, D.J.; CALLAWAY, J.C.; BRITO, G. S.; NEVES, E.S.; OBERLENDER, G.; SAIDE, O.L.; LABIGALINI, E.; TACLA, C.; MIRANDA, C.T.; STRASSMAN, R.J.; BOONE, K.B. (1996). "Human psychopharmacology of hoasca, a plant hallucinogen used in ritual context in Brazil," *Journal of Nervous & Mental Disease,* no. 184, vol. 2, pp. 86-94. [Results of the Hoasca Project], available at: http://www.udv.org.br/portugues/downloads/03.pdf and http://www.udv.org.br/portugues/downloads/02.rtf.

GROISMAN, A.; SELL, A.B. (1995). "Healing power: neurophenomenology, culture and therapy of Santo Daime," in: Winkelman, M.; Andritzky, W. (eds.). *Sacred plants, consciousness, and healing. Yearbook of cross-cultural medicine and psychotherapy.* Berlin, Verlag für Wissenschaft und Bildung, pp. 241-255.

GROISMAN, A. (1998). "Ayahuasca in Europe," *MAPS Bulletin,* no. 8, vol. 3, p. 17, available at: http://www.maps.org/ news-letters/v08n3/08317gro.html.

GROISMAN, A. (2000). "Santo Daime in the Netherlands: an anthropological study of a new world religion in a European setting," PhD Thesis in Social Anthropology, University of London.

GROISMAN, A; DOBKIN DE RIOS, M. (2007). "Ayahuasca, the U.S. supreme court and the UDV-US government case: culture, religion and implications of a legal dispute," in: Winkelman, M.; Roberts; T. (eds.). *Psychedelic medicine: new evidence for hallucinogenic substances as treatments,* vol. 1, Westport, Praeger, pp. 251-269.

HERNANDEZ, A. (2006). The rapid expansion of independent Ayahuasca groups: socio-cultural implications for Brazil and the United States. Project in Development Sociology, Cornell University. (Research awaiting funding.)

HIRUKAWA, T.; HIRAOKA, R.; SILVA, F.E. da; PILATO, S.; KOKUBO, H. (2006). "Field REG experiments of religious rituals and other events in Paraná, Brazil." 30 Encontro Psi, Implicações e Aplicações da Psi. Curitiba, Unibem, pp. 17-26.

HOOD JR., R.W. (2004). Review of: The antipodes of the mind: charting the phenomenology of the Ayahuasca experience, *The International Journal for the Psychology of Religion,* no. 14, vol. 2, pp. 147-148.

HORGAN, J. (2003). "Mapping out the Ayahuasca netherworld": a review of Benny Shanon's Antipodes of the mind, *MAPS Bulletin,* no. 13, vol. 1, available at: http://www.maps.org/news-letters/v13n1/13147hor.html.

KRIPPNER, S. (1999). "Common aspects of traditional healing systems across cultures," in: Jonas, W.B.; Levin, J.S. (eds.). *Essentials of complementary and alternative medicine,* Baltimore, Lippincott Williams & Wilkins, pp. 181-199.

KRIPPNER, S.; SULLA, J. (2000). "Identifying spiritual content in reports from Ayahuasca sessions," *International Journal of Transpersonal Studies,* no. 19, pp. 59-76.

KÜRBER, S.; RABE, D.; SONCZYK, B.; van ALEBEEK, R. (2007). "The Santo Daime church. The protection of freedom of religion under international law," The International Law Clinic of the University of Amsterdam.

LABATE, B.C.; SANTOS, R.G. dos; ROSE, I.S. DE; DERIX, G. (2007). "Bibliography of the Brazilian Ayahuasca religions," *Multidisciplinary Association for Psychedelic Studies Bulletin,* no. 17, vol. 1, pp. 28-29, available at: http://www.maps.org/news-letters/v17n1/maps-sprsmr07.pdf.

Above: *Banisteriopsis caapi* vine
Photo by Denizar Missawa Camurça

Below: Fruits of the *Psychotria viridis* shrub
Photo by Débora Carvalho Pereira Gabrich

LABIGALINI JR., E.; DUNN, J. (1995). "The Union of the Vegetable: the ritualized use of hoasca tea," *Psychiatric Bulletin,* no. 19, pp. 313-314. [General commentary on the UDV and the Hoasca Project]

LÁZARO, A. (1994). "The light in the doctrine Saint Daime," II Congresso Internacional de Estudos dos Estados Modificados de Consciência, Barcelona. (Manuscript)

LIMA, F.A. de S. (1996-1997). "The ritual use of Hoasca: comments and advice," *MAPS Bulletin,* no. 7, vol. 1, pp. 25-26, available at: http://www.maps.org/news-letters/v07n1/07125udv.html. [Recommendations and precautions for the use of ayahuasca from the Medical-Scientific Department of the UDV]

LUNA, L.E. (2006). "Traditional and syncretic Ayahuasca rituals," in: Jungaberle, H.; Verres, R.; DuBois, F. (eds.). *Rituale erneuern. Ritualdynamik und grenzerfahrung aus interdisziplinärer perspektive.* Giessen, Psychosozial-Verlag, pp. 319-338.

LUNA, L.E.; WHITE, S. (eds.) (2000). *Ayahuasca reader. encounters with the Amazon's sacred vine.* Santa Fe, NM, Synergetic Press.

LUNA, L.E.; WHITE, S. (2000). "Hyms received by Raimundo Irineu Serra, Sebastião Mota de Melo, Daniel Pereira de Matos and Francisca Campos do Nascimento," in: Luna, L.E.; White, S. (eds.). *Ayahuasca reader. encounters with the Amazon's sacred vine.* Santa Fe, NM, Synergetic Press, pp. 135-144.

MacRAE, E. (1998). "Santo Daime and Santa Maria: the licit ritual use of Ayahuasca and the illicit use of cannabis in a Brazilian Amazonian religion," *International Journal of Drug Policy,* no. 9, pp. 325-338.

MacRAE, E. (1999). "The ritual and religious use of Ayahuasca in contemporary Brazil," in: Taylor, W.; Stewart, R.; Hopkins, K.; Ehlers, S. (eds.). *DPF XII Policy Manual.* The Drug Policy Foundation Press, Washington, pp. 47-50.

MacRAE, E. (2004). "The ritual use of Ayahuasca in three Brazilian religions," in: COOMBER, R.; SOUTH, N. (eds.). *Drug use and cultural contexts "Beyond the West,"* London, Free Association Books, pp. 27-45, available at: http://www.Neip.info/downloads/t_edw11.pdf.

MacRAE, E. (2006). Guided by the moon: shamanism and the ritual use of Ayahuasca in the Santo Daime religion in Brazil (online book). Núcleo de Estudos Interdisciplinares sobre Psicoativos – Neip, available at: http://www.Neip.info/downloads/t_edw2.pdf.

MAVRICK, C. (2000). "Hallucinogens and religious identity in the Brazilian Amazon," Senior Thesis Anthropology Program, Illinois State University.

McKENNA, D.J. (2000). "An unusual experience with 'hoasca': a lesson from the teacher," in: LUNA, L.E.; WHITE, S. (eds.). *Ayahuasca reader: Encounters with the Amazon's sacred vine.* Santa Fe, NM, Synergetic Press, pp. 154-157.

McKENNA, D.J.; CALLAWAY, J.C.; GROB, C.S. (1998). "The scientific investigation of Ayahuasca: a review of past and current research," *The Heffter Review of Psychedelic Research,* no. 1, pp. 65-77. [Results of the Hoasca Project], available at: http://www.neip.info/downloads/Sci-Inv%20of%20aya-HR-DJMk.pdf.

MERCANTE, M.S. (2004). "Miração and healing: a study concerning spontaneous mental imagery and healing process," Toward a Science of Consciousness Conference, Tucson. (Manuscript)

MERCANTE, M.S. (2006a). "Images of healing: spontaneous mental imagery and healing process of the Barquinha, a Brazilian Ayahuasca religious system," PhD Thesis in Social Sciences, Saybrook Graduate School and Research Center, available at: http://www.neip.info/downloads/mercante.pdf.

MERCANTE, M.S. (2006b). "The objectivity of spontaneous mental imagery: the spiritual space of a Brazilian-Amazonian religion experienced by sacramental users of Ayahuasca," Toward a Science of Consciousness Conference, Tucson. (Manuscript)

MERCANTE, M.S. (2006c). "The objectivity of spiritual experiences: spontaneous mental imagery and the spiritual space," Revista Eletrônica Informação e Cognição, no. 5, vol. 1, pp. 78-91, available at: http://www.portalppgci.marilia.unesp.br/reic.

METZNER, R. (ed.) (1999). *Ayahuasca: human consciousness and the spirits of nature.* New York, Thunder's Mouth Press.

METZNER, R. (ed.) (2005). *Sacred vine of spirits: Ayahuasca.* Rochester, VT, Park Street Press, 2nd ed.

METZNER, R. (2006). "Varieties of ritual involving altered states of consciousness," in: JUNGABERLE, H.; VERRES, R.; DUBOIS, F. (eds.). Rituale erneuern. *Ritualdynamik und grenzerfahrung aus interdisziplinärer perspektive.* Giessen, Psychosozial-Verlag, pp. 253-266.

MEYER, M. (2003). "The seringueiro, the caboclo, and the forest queen: origin narratives of an Ayahuasca church," *Núcleo de Estudos Interdisciplinares sobre Psicoativos – Neip,* available at: http://www.Neip.info/downloads/t_matthew_seringueiro.pdf.

MEYER, M. (2005). "Religious freedom on trial," *Anthropology News,* no. 46, vol. 7, p. 27.

MEYER, M. (2006a). "Religious freedom and United States drug laws: notes on the UDV-USA legal case," *Núcleo de Estudos Interdisciplinares sobre Psicoativos – Neip,* available at: http://www.Neip.info/downloads/Matthew%20UDV-USA%20case.pdf.

MEYER, M. (2006b). "Santo Daime: cultural politics and embodied experience in Brazilian Amazônia," PhD Project in Sociocultural Anthropology, University of Virginia. (Project in development)

MIRANDA, C.T.; LABIGALINI, E.; TACLA, C. (1995). "Alternative religion and outcome of alcohol dependence in Brazil," *Addiction,* no. 90, vol. 6, p. 847. [Commentary on the Hoasca Project and on research with ex-abusers of alcohol in the UDV]

NIXON, G. (2004). Review of: The antipodes of the mind: chartering the phenomenology of the Ayahuasca experience, *Journal of Consciousness Studies,* no. 11, vols. 5-6, pp. 181-183.

PÉREZ, G.L. (2003). Review of: O uso ritual da Ayahuasca. Tipiti, *Journal of The Society for the Anthropology of Lowland South America*, no. 1, vol. 1, pp. 126-130, available at: http://www.Neip.info/downloads/1_bia1_laura_ingles.htm.

PINCHBECK, D. (2006). *2012: the return of Quetzalcoatl*. New York, Penguin Books.

QUINLAN, M. (2001). Healing from the gods: Ayahuasca and the curing of disease states. PhD Thesis in Integral Studies. California Institute of Integral Studies. [Results of research on reports from five co-researchers and the author based on two visits to Céu do Mapiá/Santo Daime-Cefluris, 1996 and 1999]

RICHMAN, G.D. (1990). "The Santo Daime doctrine: an interview with Alex Polari de Alverga," *Shaman's Drum*, no. 22, pp. 30-41.

ROHDE, S.A. (2004). Review of: Wege zum heil: die Barquinha. Eine ethnologische studie zu transformation und heilung in den Ayahuasca-ritualen einer brasilianischen religion, *Archido*, available at: http://www.archido.de/index.php?option=com_content&task=view&id=31.

SANTOS, R.G. dos; LANDEIRA-FERNANDEZ, J.; STRASSMAN, R.J.; CRUZ, A.P.M. (2007). "Effects of Ayahuasca on psychometric measures of anxiety, panic-like and hopelessness in Santo Daime members," *Journal of Ethnopharmacology*, no. 112, vol. 3, pp. 507-513. [Results of research conducted with long-term consumers of ayahuasca nine members of Santo Daime with ten years' minimum experience), Brasília (DF), 2006]

SAUNDERS, N.; SAUNDERS, A; PAULI, M. (2000). "A different kind of church" in: SAUNDERS, N.; SAUNDERS, A.; PAULI, M. *In search of the ultimate high: spiritual experiences through psychoactives*. London, Rider & Co.

SCHINZINGER, A. (2001). "Mysterious tea," in: ROBERTS, T. (ed.). *Psychoactive sacramentals: essays on entheogens and religion*. San Francisco, Council on Spiritual Practices, pp. 103-110.

SCHINZINGER, A. *Liquid light: Ayahuasca, dreams and healing.* [Unpublished manuscript]

SHANON, B. (2002). *The antipodes of the mind: charting the phenomenology of the Ayahuasca experience.* Oxford, Oxford University Press. [Results of research with about 200 individuals from indigenous groups and mestizo populations in Colombia, Ecuador, and Peru, and with members of the Barquinha, the União do Vegetal, Santo Daime and of alternative groups in Brazil and Europe, between 1994 and 2000], available at: http://alto-das-estrelas.blogspot.com/2007/01/summary-of-benny-shanons-book-on.html.

SOIBELMAN, T. (1995). "My father and my mother, show me your beauty: ritual use of Ayahuasca in Rio de Janeiro," Master of Social and Cultural Anthropology, California Institute of Integral Studies, available at: http://alto-das-estrelas.blogspot.com/2006/12/dissertao-de-mestrado-sobre-uso-da.html.

STING [STEVE BORDEN] (2003). *Broken music: a memoir.* New York, The Dial Press.

SULLA, J. (2005). "The system of healing used in the Santo Daime community Céu do Mapiá," Master of Arts in Psychology, Saybrook Graduate Scholl and Research Center. 2005. [Results of research with members of the community in Céu do Mapiá/Santo Daime-Cefluris and of the author's experiences as a member of the group, 1995 to 2004]

TUPPER, K.W. "The globalization of Ayahuasca: Harm reduction or benefit maximization?" *International Journal of Drug Policy.* (Unpublished manuscript)

VAN DER PLAS, A.G. (2002). "International legal aspects of the use of Ayahuasca," Conference on Psychoactivity III, Amsterdam.

VILLAESCUSA, M. (2002). *An exploration of psychotherapeutic aspects of Santo Daime ceremonies in the UK, MSc in Humanistic Integrative Psychotherapy,* Middlesex University. [Results of research with six participants in Santo Daime-Cefluris rituals in London with at least one year's experience, 2001 to 2002], available at: http://www.Neip.info/downloads/villaescusa/MSc%20Dissertation%202003.pdf.

WATERMAN, D. Sacred medicines: entheogens, society & law. (Unpublished manuscript)

WEINHOLD, J. (2007). "Beyond pharmacology: the efficacy of rituals within the European Santo Daime religion," Paper presented at the Annual Conference ("The Efficacy of Rituals") of the Collaborative Research Centre "Ritual Dynamics," Heidelberg University.

WEINHOLD, J. (in press). "Failure and mistakes in rituals of the european Santo Daime church: experiences and subjective theories of participants," in: Hüsken, U. (ed.). *Ritual failure, mistakes in ritual, and ritual dynamics*. Leiden, Brill, available at: http://alto-das-estrelas.blogspot.com/2007/04/heidelberg-rearch-iv.html.

WEISKOPF, J. (2004). *Yajé, the new Purgatory*. Bogotá, Villegas Editores, available at: http://alto-das-estrelas.blogspot.com/2005/05/livro-de-jimmy-weiskopf-sobre-o-yag-na.html.

WEISKOPF, J. (2006). Review of: O uso ritual da Ayahuasca. Erowid, available at: http://www.erowid.org/library/review/review.php?p=191 and http://www.neip.info/downloads/ resenha %20Jimmy%20Weiskopf%20ingl%EAs.htm.

French

BIZE, C. (2003). "Ayahuasca: nouveaux usages d'une drogue ancienne," Thèse de pharmacie. Université Paris 11.

BOIS-MARIAGE, F. (2002). "Ayahuasca: une synthèse interdisciplinaire," *Psychotropes*, no. 8, vol. 1, pp. 79-113.

CHAUMEIL, J.-P. (1992). "Un nouveau testament pour un troisième millénaire: la religion du Daime au Brésil," *Cahiers d'Études africaines*, no. 125, pp. 151-155.

DESHAYES, P. (2002). "L'ayawaska n'est pas un hallucinogène," *Psychotropes*, no. 8, vol. 1, pp. 65-78.

DESHAYES, P. (2004). "De l'amer à la mère: quiproquos linguistiques autour de l'Ayahuasca," *Psychotropes*, no. 10, vols. 3-4, pp. 15-29.

DESHAYES, P. (2006). "Les trois mondes du Santo Daime," *Socio-anthropologie*, no. 17-18, pp. 61-83, available at: http://socio-anthropologie.revues.org/document451.html.

LAMOUREUX, P. (2001). "Le Santo Daime à Madrid," Mémoire de Maîtrise d'Ethnologie, Université Paris 7.

MACRAE, E. (1998). "L'utilisation religieuse de l'Ayahuasca dans de Brésil contemporain," *Cahiers du Brésil Contemporain*, no. 35-36, pp. 247-254.

MARTINS-VELLOSO, S. (2003). "Imaginaire collectif, productions oniriques et modalités de consommation de l'Ayahuasca en forêt amazonienne," *Mémoire de fin de Capacité en Addictologie Clinique*, Université Paris 5.

NOVAES, C. (2006). "L'expérience de l'Ayahuasca et ses états modifiés de la conscience. Une étude transculturelle des récits des usagers urbains de l'Ayahuasca. Une lecture a travers le concept de l'inconscient selon Gilles Deleuze," Master recherche "Développement, Psychopathologie et Psychanalyse, Clinique Transculturelle." Option Clinique Transculturelle. Université Paris 13. [Results of research with three members of do Santo Daime in São Paulo collected on the internet, with two members of the União do Vegetal in Londrina (Paraná state), and seven members of "alternative groups" in Assis (São Paulo state), collected on the internet, 2004-2006], available at: http://www.Neip.info/downloads/clara_fr/clara_fr_tese.pdf.

PERLONGHER, N. (1990). "La force de la forme. Notes sur la religion du Santo Daime," *Societés*, no. 2, pp. 21-30, available at: http://www.Neip.info/downloads/t_perlongher.pdf.

German

ADELAARS, A; CLAUDIA, M.-E.; RÄTSCH, C. (2006). *Ayahuasca, zaubertraenke, rituale und kunst aus Amazonien*. Aarau, AT Verlag. 2006. http://alto-das-estrelas.blogspot.com/2007/01/novo-livro-em-alemo-sobre-Ayahuasca-e.html.

ADELAARS, A. (2006). "Santo Daime: die bekannteste brasilianishe Ayahuasca-kirche," in: Adelaars, A; Claudia, M.-E.; RÄTSCH, C. *Ayahuasca, zaubertraenke, rituale und kunst aus Amazonien*. Aarau, AT Verlag, pp. 231-245.

ALVERGA, A.P. de (2007). *Der prophet aus dem regenwald: begegnung mit heiligen pflanzen.* Burgrain, KOHA-Verlag GmbH.

AUF DEM HÖVEL, J. (2005). "Droge oder sakrament? Ayahuasca kommt vom brasilianischen in den Großstadt-Dschungel." Heise, available at: http://www.heise.de/tp/r4/artikel/19/19285/1.html. (Original article published in: *HanfBlat,* 94. 2005)

BAGULEY, E. (2006). "Das singende Volk von Juramidam: lieder der brasilianischen und europäischen Santo Daime-religion," Magisterarbeit im Fach Völkerkunde, Universität Philipps-Marburg, available at: http://alto-das-estrelas.blogspot.com/2006/06/tese-de-antropologia-em-alemo-sobre.html.

BAIKER, D. (2004). "Wanderung einer religion: der gebrauch von halluzinogenen in der Santo-Daime-kirche in Brasilien und Europa," Magisterarbeit im Fach Völkerkunde, Universität Philipps-Marburg.

BALZER, C. (1998). "Wege zum heil: die Barquinha. Ein religiöses rettungsboot auf wogen des kulturellen und sozialen chaosmos amazonischer welten" (Amazonische transformationen im lichete Ayahuascas), Magisterarbeit in Etnologie, Freie Universität Berlin.

BALZER, C. (1999). "Santo Daime in Deutschland: eine verbotene Frucht aus Brasilien," *Zeitschrift für Religionswissenschaft,* no. 7, pp. 49-79.

BALZER, C. (2003). "Wege zum heil: die Barquinha. Eine ethnologische studie zu transformation und heilung in den Ayahuasca-ritualen einer brasilianischen religion," Mettingen, Brasilienkunde-Verlag, available at: http://alto-das-estrelas.blogspot.com/ 2007/01/book-about-barquinha-in-german.html.

DERIX, G. (2004). "Ayahuasca. Eine Kritik der psychedelischen Vernunft. Philosophisches abenteuer am amazonas," Solothurn, Nachtschatten Verlag, available at: http://alto-das-estrelas.blogspot.com/2006/12/livro-em-alemo-e-holands-sobre.html and http://alto-das-estrelas.blogspot.com/2006/12/summary-in-english-of-govert-derixs.html.

FIEDLER, L. (2007). "Struktur und geschichte von Ayahuasca: and Santo

Daime: ritualen unter berücksichtigung medizinischer aspekte," Medical Doctor project at Medizinische Fakultät der Universität Heidelberg. (Project in development) [Results of research with eighteen individuals from Germany, the Netherlands, and Switzerland, among them thirteen members of Santo Daime churches and five participants from other ayahuasca groups led by South American shamans or by their European disciples or Europeans guided by therapeutic paradigms, between 2003 and 2006], available at: http://alto-das-estrelas.blogspot.com/2007/03/heidelberg-research-ii-lisa-fiedler.html.

FISCHER-F., R. (1996). "Fliegender pfeil: eine frau folgt dem ruf des Ayahuasca in den dschungel," Munich, Wilhelm Heyne Verlag, available at: http://alto-das-estrelas.blogspot.com/2007/01/livro-em-alemo-sobre-o-santo-daime.html.

HUTTNER, J. (1998). "Santo Daime: eine neue heilsbewegung," Magisterarbeit in Antropologie, Johann Wolfgang Goethe Universität, available at: http://alto-das-estrelas.blogspot.com/2007/03/master-thesis-in-anthropology-about.html.

JUNGABERLE, H. (2006). "Rituale und integrationskompetenz beim gebrauch psychoaktiver substanzen," in: Jungaberle, H.; Verres, R.; DuBois, F. (eds.). *Rituale erneuern. Ritualdynamik und grenzerfahrung aus interdisziplinärer perspektive*. Giessen, Psychosozial-Verlag, pp. 77-106.

JUNGABERLE, H. (2007). "Ritualtransfer und die soziopharmakologie des drogengebrauchs," in: Ahn, G.; Robert, L.; Snoek, J. (eds.). *Ritualtransfer*. Amsterdam, Rio de Janeiro, Strawberry Plains Tennessee. (in press)

JUNGABERLE, H.; VERRES, R.; DUBOIS, F. (eds.). (2006). *Rituale erneuern. Ritualdynamik und grenzerfahrung aus interdisziplinärer perspektive*. Giessen, Psychosozial-Verlag, available at: http://alto-das-estrelas.blogspot.com/2007/01/lanado-novo-livro-em-alemo-sobre-o-uso.html.

JUNGABERLE, H.; LUNA, L.E.; METZNER, R.; STYK, J.; WECKER, K. (2006). "Können rituale drogenmissbrauch verhindern?" – Diskussion über Rituale und Drogengebrauch, in: J JUNGABERLE, H.; VERRES, R.; DUBOIS, F.

(eds.). *Rituale erneuern. Ritualdynamik und grenzerfahrung aus interdisziplinärer perspektive.* Giessen, Psychosozial-Verlag, pp. 357-365.

LONGI, A. (2000). "Ayahuasca: vom Amazonas in die Dschungel der Städte," *Pharmakeia,* available at: http://www.pharmakeia.com/santodaimeaffaire.htm.

MEYERRATKEN, U.; SALEM, N. (1998). *Daime: Brasiliens kult der heilenden kraftpflanzen.* Munich, Knaur, available at: http://alto-das-estrelas.blogspot.com/2007/01/book-about-santo-daime-in-german.html.

ROHDE, S.A. (1996). "Erste orientierung über rituellen Ayahuasca-gebrauch in ausgewählten stammeskulturellen arealen und sozioökonomischen Umwelten Südamerikas," *Pharmakeia,* available at: http://www.pharmakeia.com/Ayahuasca.htm.

ROHDE, S.A. (1997). "Aspekte rituellen verhaltens am beispiel eines heilrituals des brasilianischen Santo Daime kults," *Pharmakeia,* available at: http://www.pharmakeia.com/heilritu.html.

ROHDE, S.A. (2001). "Formen des religiösen gebrauches des entheogenen sakramentes Ayahuasca unter besonderer berücksichtigung des aspektes ihrer kriminalisierung," *Diplomarbeit im Fach Religionswissenschaft,* Universität Bremen.

ROHDE, S.A. (2004). "Santo Daime: die religion der königin des waldes (Teil 1)," *Entheogene Blätter,* no. 20, pp. 51-56.

ROHDE, S.A. (2004). "Santo Daime: die religion der königin des waldes (Teil 1)," *Entheogene Blätter,* no. 21, pp. 92-95.

ROHDE, S.A. (2006). "Rezension: o uso ritual da Ayahuasca," *Archido,* available at: http://www.archido.de/index.php?option=com_content&task=view&id=25&Itemid=12 and http://www.neip.info/downloads/resenha_usoritual_alema.pdf.

SCHMID, J. (2007). "Droge oder medizin? Selbstbehandlungsversuche im kontext von Ayahusca: oder Santo Daime ritualen. Gesundheitskonzepte im umfeld des ritualisierten drogengebrauchs in Europa. Eine qualitative studie zur ritualdynamik und salutogenese," Doctorum scienciarum humanarum project at Medizinische Fakultät der Universität Heidelberg (Project in development) [Results of research with sixteen individuals from Germany, Austria, Denmark, Spain, England, the Netherlands, and Switzerland, ten of them members of Santo Daime churches and six participants from other ayahuasca groups led by South American shamans, by their European disciples or by Europeans guided by therapeutic paradigms, as well as individuals who consume ayahuasca at home or independent drinkers who frequent different contexts, between 2003 and 2006], available at: http://alto-das-estrelas.blogspot.com/2007/03/heidelberg-research-iii.html.

SIEBERT, U. (1998). "Angstreduktion durch Santo Daime. Ethnomedizinische untersuchungen im brasilianischen regenwald." *Curare, Zeitschrift dür Ethnomedizin und Transkulturelle Psychiatrie,* no. 14, pp. 149-154.

Italian

BIANCHI, A. (2004). Review of: O uso ritual da Ayahuasca. Organizzazione interdiciplinare sviluppo e sallute – Oriss, available at: http://www.oriss.org/articoli/a_Ayahuasca.html and http://www.Neip. info/downloads/l_bia1_antonio_bianchi.htm.

CURUCHICH, T.; CRUZ O. (n.d.). *Chiesa regina della pace* (Cielo d'Assisi). Bienni di Specializzazione in Teologia Fondamentale, Istituto Teológico di Assisi.

GIOIA, W. (1996). *Alle orgenti dell'Essere. Viaggio interiore nella foresta amazzonica.* Pádova, Casa Editrice Meb, available at: http://alto-das-estrelas.blogspot.com/2007/01/livro-em-italiano-sobre-o-santo-daime_29.html.

INTROVIGNE, M. (2000). "Le Chiese del Santo Daime dal Brasile all'Europa: tra profezia e polizia." *Center for Studies on New Religions – Cesnur,* available at: http://www.cesnur.org/testi/mi_daime2K.htm.

KOLIOPOULOS, N. (2006). "L'esistenza giuridica delle chiese enteogeniche," Tesi de Laurea en Antropologia Giuridica, Università degli Studi di Macerata.

LABATE, B.C. (2003). Piante che curano, in: BIANCHI, A.; COZZI, L.; SPECIALE, A. (eds.). *Forest medicine: la medicina tradizionale una risorsa da conoscere*. Milão, Centro Orientamento Educativo, pp. 87-90.

LUCÀ T.G. (1995). "Sincretismo religioso e identità. Riflessioni a partire da una ricerca in Brasile," *Religione e Società*, no. 21, pp. 104-125.

LUPPICHINI, E. (2006). "Dottrina e pratica del Santo Daime. Una forma di nuova spiritualità tra Amazzonia ed Europa," Tesi di Laurea in Antropologia Culturale, Università degli Studi di Pisa, available at: http://www.neip.info/downloads/elena/tese_elena.pdf.

MENOZZI, W. (2007). *Ayahuasca, la liana degli spiriti – il sacramento magico-religioso dello sciamanismo amazzonico*. Milão, Franco Angeli Editore.

PALMIERI, M. (2006). "L'incontro col santo Daime," *Altrove*, 12: 99-112.

PRASCINA, A. (1997). "Origine e sviluppo del culto amazzonico del Santo Daime," Tesi di Laurea in Psicologia, Università degli Studi di Padova. [Results of research conducted at the Colônia Cinco Mil and in Céu do Mapiá, both groups of Santo Daime-Cefluris, during the December festivals, 1996-1997]

VENECE, C. (2004). "Riti di guarigione e iniziazione nei culti sincretici moderni," *Altrove*, no. 10, pp. 113-128.

VERLANGIERI, A. (1998). "Ayahuasca: un fenomeno sciamanico per il terzo millennio," *Altrove*, no. 5, pp. 119-134.

VERLANGIERI, A. (2000). *Il maestro della foresta. Santo Daime: un'esperienza sciamanica*. Murazzano, Ellin Selae, available at: http://alto-das-estrelas.blogspot.com/2007/02/livro-sobre-o-santo-daime-em-italiano.html.

Japanese

AKIRA (2001). *Ayawasuka!: chijo caikyo no drug o motomete*. Tokyo, Kodansha.

HIRUKAWA, T. (2002). *Higan no jikan: ishiki no jinruigaku*. Tokyo, Shunjusha.

NAKAMAKI, H. (1992a). "Cha o nomazunba genkaku o ezu," in: WAKIM-

OTO, H.; YANAGAWA, K. (eds.). *Gendai shukyogaku 1: shukyo taiken e no sekkin,* Tokyo, Tokyo University Press, pp. 31-59. [Translation of the article, originally published in Portuguese, "Quem não toma o chá não tem alucinações: epidemiologia de religiões alucinógenas no Brasil"]

NAKAMAKI, H. (1992b). "Hajime ni cha ariki: brazil ni okeru genkaku shukyo no soseiki," in: NAKAMAKI, H. (ed.). *Tosui suru bunka.* Tokyo, Heibonsha, pp. 17-49.

NAKAMAKI, H. (1997). "Mirason to buhasheira," *Study of Transpersonal,* no. 2, pp. 28-39.

NAKAMAKI, H. (2001). "Amazon no shamanic vision no hyosho," Kokuritsu Minzokugaku Hakubutsukan Kenkyu Houkoku Bessatsu 22 go, pp. 113-124.

NAGATAKE, H. (1995). *Magical herb: nanbei no genkakusei shokubutsu to shamanic healer.* Tokyo, Daisan Shokan.

Norwegian

DAMN, I. (1999). "Til te hos regnskogens guder: brasiliansk kult byr på sterk trylledrikk," *Illustrert vitenskap,* no. 8.

Portuguese

A ARCA DA UNIÃO. Online magazine. Rio Branco, available at: http://www.arcadauniao.org. jun/05-set/06. [Online magazine specializing in themes related to Santo Daime, but also in shamanism and the ritualized use of psychoactive substances]

ABRAMOVITZ, R.S. de M. (2002). "Música e silêncio na concentração do Santo Daime," *Cadernos do Colóquio,* no. 1, vol. 5, pp. 20-29.

ABRAMOVITZ, R.S. de M. (2003). "Música e miração: uma análise etnomusicológica dos hinos do Santo Daime," Mestrado em Música. UNI-RIO, available at: http://alto-das-estrelas.blogspot.com/2006/02/tese-de-mestrado-faz-anlise.html.

ABREU, R.M. do R.M. de (1984). "Daime Santa Maria: uma antropologia de áudios e imagens," *Comunicações do Iser,* no. 10, vol. 1, pp. 36-46.

ABREU, R.M. do R.M. de (1990). "A doutrina do Santo Daime," in: LANDIM, L. (ed.). *Sinais dos tempos*. Rio de Janeiro, Instituto de Estudos da Religião, pp. 253-263.

AFONSO, C.A. (2007). "Paródia sacra," *Núcleo de Estudos Interdisciplinares sobre Psicoativos – Neip*, available at: http://www.Neip.info/downloads/c_afonso/parodia_sacra.pdf.

ALBUQUERQUE, M.B.B. (2006). "Pedagogia da floresta: um estudo sobre práticas educativas centradas no culto ao Santo Daime," Anais do VI Congresso Luso Brasileiro de História da Educação. Percursos e Desafios da pesquisa e do ensino de história da Educação. Uberlândia, pp. 4924-4934.

ALBUQUERQUE, M.B.B. (n.d.). *Máximas do Padrinho Sebastião*. (Manuscript)

ALBUQUERQUE, M.B.B. (2007a). "Filosofia, educação e religião. Conexões a partir do Santo Daime," Anais do 18° Encontro de Pesquisa Educacional do Norte e Nordeste. Maceió, available at: http://www.Neip.info/downloads/maria_betania_religiao.pdf.

ALBUQUERQUE, M.B.B. (2007b). "Filosofia, religião e educação: conexões a partir do Santo Daime," Comunicação no 180 Epenn: Encontro de Pesquisa Educacional do Norte e Nordeste. Maceió.

ALBUQUERQUE, M.B.B. (2007c). *ABC do Santo Daime*. Belém, Eduepa, available at: http://alto-das-estrelas.blogspot.com/2007/11/lanado-livro-sobre-o-santo-daime.html.

ALBUQUERQUE, M.B.B.; COLARES, N.J.; BORGES, K.C.R.; VIANA, R.L.; BARBOSA, K.J.C. (2006). "Práticas educativas da religião do Santo Daime," III Fórum de Pesquisa, Ensino, Extensão e Pós-Graduação da Uepa. Educação, Cultura, Identidade e Diversidade Amazônica, Belém.

ALMEIDA, F.; FIGUEIREDO, E.; DEUS, J. de (1996). *Mestre Antonio Geraldo e o Santo Daime. Centro Espírita Daniel Pereira de Matos Barquinha*. Rio Branco, Bobgraf.

ALMEIDA, M. de F.H. de (2002). "Santo Daime: a colônia cinco mil e a contracultura (1977-1983)," Mestrado em História, UFPE.

ALVERGA, A.P. de (1984a). *O livro das mirações: viagem ao Santo Daime*. Rio de Janeiro, Editora Rocco.

ALVERGA, A.P. de (1984b). *O guia da floresta*. Rio de Janeiro, Editora Nova Era/Record.

ALVERGA, A.P. de (1992). *O Guia da floresta*. 2nd ed., Rio de Janeiro, Editora Record.

ALVERGA, A.P. de (1995). *O livro das mirações: viagem ao Santo Daime*. 2nd ed., Rio de Janeiro, Editora Record/Nova Era.

ALVERGA, A.P. de (1996). "Seriam os deuses alcalóides?," Texto apresentado no Congresso Internacional de Psicologia Transpessoal, Manaus.

ALVERGA, A.P. de (1998). *O evangelho segundo Sebastião Mota*. 2nd ed., Céu do Mapiá, Cefluris Editorial.

ALVES, A.M. (2005). "Tambores para a rainha da floresta: a inserção da umbanda no Santo Daime," Mestrado em Ciências da Religião, PUC-SP, available at: http://alto-das-estrelas.blogspot.com/2007/03/pesquisa-investiga-relao-do-santo-daime.html.

ALVES, A.M. (2007). "Tambores para a rainha da floresta: a inserção da umbanda no Santo Daime," XIV Jornadas Sobre Alternativas Religiosas na América Latina, Buenos Aires, Unsam. Available at http://www.neip.info/downloads/ge/ ge_marques.pdf.

ANDRADE, A.P. de (1995a). "O fenômeno do chá e a religiosidade cabocla," Mestrado em Ciência das Religiões, Instituto Metodista de Ensino Superior, available at: http://www.neip.info/downloads/afranio/afranio_01.pdf.

ANDRADE, A.P. de (1995b). "O fenômeno do chá e a religiosidade cabocla: um estudo centrado na União do Vegetal," in: Cetad/UFBa (eds.). I Encontro de Estudos sobre Rituais Religiosos e Sociais e o Uso de Plantas Psicoativas, Seminário Internacional: O Uso e o Abuso de Drogas (Programa Oficial). Salvador, Cetad/UFBa.

ANDRADE, A.P. de (1998). "A União do Vegetal no astral superior," *Comunica-*

ções do ISER, no. 30, pp. 61-65.

ANDRADE, A.P. de (2004). "Contribuições e limites da União do Vegetal para a nova consciência religiosa," in: LABATE, B.C.; ARAÚJO, W.S. (eds.). *O uso ritual da Ayahuasca*. 2nd ed. Campinas, Mercado de Letras, pp. 589-613.

ANDRADE, E.N.; BRITO, G.S.; ANDRADE, E.O.; NEVES, E.S.; McKENNA, D.; CAVALCANTE, J.W.; OKIMURA, L.; GROB, C.; CALLAWAY, J.C. (2004). "Farmacologia humana da Hoasca: estudos clínicos (avaliação clínica comparativa entre usuários do chá hoasca por longo prazo e controles; avaliação fisiológica dos efeitos agudos pós-ingestão do chá hoasca)," in: LABATE, B.C.; ARAÚJO, W.S. (eds.). *O uso ritual da Ayahuasca*. 2nd ed. Campinas, Mercado de Letras, pp. 671-709. [Results of the Hoasca Project]

ANDRADE, G.B. (2001). "O Ministério Público como órgão promovedor de ações que visem preservar os direitos difusos e coletivos, especificamente sobre o uso do chá vegetal denominado Santo Daime, a fim de proteger as crianças, adolescentes e doentes mentais," Congresso Nacional do Ministério Público. Recife.

ANDRADE, J. de (1981). "Música e dança na 'Miração' do Santo Daime," in: Bispo, A.A. et alli. *Musices Aptatio: Anuário de Estudos Hinológicos e Musicológicos*. Köln, Institut für hymnologische und musikethnologische Studien, pp. 299-313.

APOLÔNIO, V.; FIGUEIREDO, E. (1996). *Mestre Antonio Geraldo e o Santo Daime*. Rio Branco, Preview.

ARAÚJO, J.R. (1996). "A oaska no neo-liberalismo: o caso do Santo Daime," XIX INTERCOM, São Paulo.

ARAÚJO, J.R. (1998). "Ayahuaska e preconceito: o olhar positivista na comunicação do Matutu," III Colóquio Brasil-França de Pesquisadores da Comunicação. Sergipe, pp. 303-316.

ARAÚJO, J.R. (1999). "O discurso do xamã e do lugar de onde fala: estudo de comunicação e mediações em culturas excluídas," Doutorado em Ciências da Comunicação, USP.

ARAÚJO, J.R. (2002). "Comunicação e exclusão: a leitura dos xamãs," São Paulo, Arte & Ciência.

ARAÚJO, J.R. (2007). "Hospitalidade, cura e Santo Daime: um estudo de caso," *Núcleo de Estudos Interdisciplinares sobre Psicoativos – Neip,* available at: http://www.neip.info/downloads/jussara/jussara_01.pdf.

ARAÚJO, M.C.R.; CASTRO, R.V. (2004). "Estados alterados de consciência e produção de subjetividade: a miração daimista pensada como virtualidade," X Encontro da Associação Brasileira de Psicología Social (Abraspso), Regional Sul. A Psicologia social em movimento e a realidade brasileira: por onde e para onde caminhamos. Curitiba.

ARAÚJO, M.C.R. (2005a). "Partilhando idéias: Santo Daime: teoecologia e adaptação aos tempos modernos," *Núcleo de Estudos Interdisciplinares sobre Psicoativos – Neip,* available at: http://www.neip.info/mainwebsite_html/downloads/textos%20novos/Texto%20Maria%20Clara.pdf.

ARAÚJO, M.C.R. (2005b). "Santo Daime: teoecologia a adaptação aos tempos modernos," Mestrado em Psicologia Social, Uerj. [Results of research with fifteen members with at least ten years' experience in several Santo Daime-Cefluris churches in Rio de Janeiro, 2003 to 2005]

ARAÚJO, M.C.R.; CASTRO, R.V. (2005). "A miração daimista e o virtual: correlações e diferenças," *Revista Último Andar: caderno de pesquisa em Ciências da Religião,* no. 15, pp. 77-94.

ARAÚJO, M.C.R.; CASTRO, R.V. (2005). Review of: A reinvenção do uso da ayahausca nos centros urbanos. *Revista de Psicologia da UnC,* no. 2, vol. 2, pp. 134-135, available at: http://www.neip.info/downloads/clara/ resenha _clara_01.pdf.

ARAÚJO, M.G.J. (1998). "Entre almas, encantes e cipó," Mestrado em Antropologia Social, Unicamp.

ARAÚJO, M.G.J. (2004). "Cipó e imaginário entre seringueiros do Alto Juruá," *Rever: Revista de Estudos da Religião,* no. 4, vol. 1, pp. 41-59.

ARAÚJO, W.S. (1996). "Sobre o balanço do mar," XX Reunião Brasileira de Antropologia, Salvador.

ARAÚJO, W.S. (1997). "Navegando sobre as águas do mar sagrado: história, cosmologia e ritual no centro espírita e culto de oração Casa de Jesus Fonte de Luz," Mestrado em Antropologia Social, Unicamp.

ARAÚJO, W.S. (1997a). "Navegar é preciso: o embarque dos marinheiros do mar sagrado na barca Santa Cruz," I Encontro de História Oral da Região Norte, Rio Branco.

ARAÚJO, W.S. (1997b). "Cosmologia em construção," I Congresso sobre o Uso Ritual da Ayahuasca (I Cura), Campinas.

ARAÚJO, W.S. (1997c). "O repertório simbólico da Barquinha," Anais do VI Encontro de Antropólogos do Norte/Nordeste. Recife.

ARAÚJO, W.S. (1999a). *Navegando sobre as ondas do Daime: história, cosmologia e ritual da Barquinha.* Campinas, Editora da Unicamp.

ARAÚJO, W.S. (1999b). "A questão do sincretismo em religiões e grupos ayahuasqueiros," VI Reunião Regional de Antropólogos do Norte e do Nordeste, Belém.

ARAÚJO, W.S. (1999c). "Categorias de transe: miração, possessão e irradiação em uma religião usuária de Ayahuasca," III Reunião de Antropologia do Mercosul, Missiones, Posadas.

ARAÚJO, W.S. (1999d). "O bailado da Barquinha: notas acerca de lazer e trabalho em uma religião usuária de Ayahuasca," IX Jornadas Alternativas Religiosas para a América Latina, Rio de Janeiro.

ARAÚJO, W.S. (1999e). "A Barquinha: uma cosmologia amazônica em construção," I Congresso de História das Religiões, Unesp/Assis.

ARAÚJO, W.S. (2002). "A questão do bem e do mal em uma religião usuária de Ayahuasca da Amazônia Ocidental," Congresso da Associação Latino/Americana de Estudos de Religião, Lima.

ARAÚJO, W.S. (2004). "Barquinha: espaço simbólico de uma cosmologia em construção," in: LABATE, B.C.; ARAÚJO, W.S. (eds.). *O uso ritual da Ayahuasca*. 2nd ed., Campinas, Mercado de Letras, pp. 541-555.

ARAÚJO, W.S. (2006). *Táticas e estratégias em uma religião usuária de Ayahuasca da Amazônia Ocidental*. Rio Branco, Uninorte.

ARRUDA, C.; LAPIETRA, F.; SANTANA, R.J. (2006). *Centro livre: ecletismo cultural no Santo Daime*. São Paulo, Editora All Print, available at: http://alto-das-estrelas.blogspot.com/2006/09/lanamento-de-livro-reportagem-sobre.html; http://alto-das-estrelas.blogspot.com/2006/09/resenha-sobre-o-livro-reportagem.html and http://alto-das-estrelas.blogspot.com/2006/09/trecho-do-livro-centro-livre-ecletismo.html.

BALZER, C. (2004). "Santo Daime na Alemanha: uma fruta proibida do Brasil no 'mercado das religiões'," in: Labate, B.C.; Araújo, W.S. (eds.). *O uso ritual da Ayahuasca*. 2nd ed. Campinas, Mercado de Letras, pp. 507-537.

BARBOSA, P.C.R. (2001). "Psiquiatria cultural do uso ritualizado de um alucinógeno no contexto urbano: uma investigação dos estados de consciência induzidos pela ingestão da Ayahuasca no Santo Daime e União do Vegetal em moradores de São Paulo," Mestrado em Saúde Mental, Unicamp. [Results of research with first-time users of ayahuasca nineteen people in Santo Daime, in São Paulo; nine people in the UDV, in Campinas and São Paulo), 1997 and 2000], available at: http://www.neip.info/downloads/paulo_tse/tese_paulo.pdf.

BARBOSA, P.C.R. (2003). "Emoção e cognição na experiência com o uso ritual da Ayahuasca no culto do Santo Daime," VII Abanne: Reunião de Antropólogos do Norte e Nordeste, São Luís.

BARBOSA, P.C.R. (2004). "Usos rituais da Ayahuasca no contexto urbano e itinerários religiosos/terapêutico," XXII Congresso Brasileiro de Psiquiatria, Salvador.

BARBOSA, P.C.R. (2006). "Uma avaliação neuropsicológica longitudinal de sujeitos que usam o alucinógeno amazônico-ameríndio Ayahuasca em um contexto religioso-ritual urbano," Qualificação de Doutorado em Saúde

Mental, Unicamp. (Project in development) [Continuation of research with first-time users of ayahuasca nineteen people in Santo Daime, in São Paulo; nine people in the UDV, in Campinas and São Paulo), 1997 and 2000]

BARBOSA, P.C.R.; DALGALARRONDO, P. (2003). "O uso ritual de um alucinógeno no contexto urbano: estados alterados de consciência e efeitos em curto prazo induzidos pela primeira experiência com a Ayahuasca," Jornal Brasileiro de Psiquiatria, no. 52, vol. 3, pp. 181-190. [Results of research with first-time users of ayahuasca nineteen people in Santo Daime, in São Paulo; nine people in the UDV, in Campinas and São Paulo), 1997 and 2000]

BASTOS, A. (1979). "Culto do Santo Daime," in: BASTOS, A. *Os cultos mágico-religiosos no Brasil*. São Paulo, Hucitec, pp. 215-217.

BOLSANELLO, D.P. (1995). *Busca do Graal Brasileiro: a doutrina do Santo Daime*. Rio de Janeiro, Bertrand Brasil.

BOMFIM, J.D. (2006). *O Jardim de Belas Flores. O Hinário O Cruzeiro Universal do Mestre Raimundo Irineu Serra comentado por Juarez Duarte Bomfim*. Online book. Centro de Iluminação Cristã Luz Universal de Minas Gerais (Ciclumig), available at: http://www.mestreirineu.org/liberdade.htm.

BORGES, F.C. (2004). "Comunicação, memória e alteridade: o caráter antropófago no movimento religioso do Santo Daime," XXVII Congresso Brasileiro de Ciências da Comunicação, Porto Alegre, available at: http://reposcom.portcom.intercom.org.br/bitstream/1904/18259/1/R0826-1.pdf.

BRANDÃO, C.R. (2004). "Fronteira da fé: alguns sistemas de sentido, crenças e religiões no Brasil de hoje," *Estudos Avançados*, no. 18, vol. 52, pp. 261-288.

BRANDÃO, P.C. (2005). "Diagnóstico geoambiental e planejamento do uso do espaço na FLONA do Purus, Amazônia Ocidental: um subsídio ao Plano de Manejo," Mestrado em Ciência Florestal. UFV, available at: http://alto-das-estrelas.blogspot.com/2006/02/tese-de-mestrado-apresenta-diagnstico.html e http://alto-das-estrelas.blogspot.com/2007/03/site-do-Neip-publica-dissertao-de.html and http://www.Neip.info/ downloads/pedro_cristo/

tese_pedro.pdf.

BRISSAC, S.G.T. (1999a). "A Estrela do Norte iluminando até o Sul: uma etnografia da União do Vegetal em um contexto urbano," Mestrado em Antropologia Social, Museu Nacional/UFRJ.

BRISSAC, S.G.T. (1999b). "Alcançar o alto das cordilheiras: a vivência mística de discípulos urbanos da União do Vegetal," IX Jornadas sobre Alternativas Religiosas na América Latina. Seminário Temático no. 5: Mística, transe e possessão: olhares sociológicos. Rio de Janeiro.

BRISSAC, S.G.T. (1999c). "José Gabriel da Costa: trajetória de um brasileiro, mestre e autor da União do Vegetal," Papers do I Simpósio da Associação Brasileira de História das Religiões, Assis.

BRISSAC, S.G.T. (1999d). "Constituição histórica da União do Vegetal," Anais do III Encontro de Pesquisa dos Estudantes de História, Campinas.

BRISSAC, S.G.T. (2004). "José Gabriel da Costa: trajetória de um brasileiro, mestre e Autor da União do Vegetal," in: LABATE, B.C.; ARAÚJO, W.S. (eds.). *O uso ritual da Ayahuasca*. 2nd ed. Campinas, Mercado de Letras, pp. 571-587.

BRITO, G. de S. (2004). "Farmacologia humana da Hoasca: chá preparado de plantas alucinógenas usado em contexto ritual no Brasil," in: LABATE, B.C.; Araújo, W.S. (eds.). *O uso ritual da Ayahuasca*. 2nd ed., Campinas, Mercado de Letras, pp. 623-651. [Results of the Hoasca Project]

CAMARGO, I.A. (1995). "Uso religioso do chá Ayahuasca," in: Cetad/UFBa (eds.). *I Encontro de Estudos sobre Rituais Religiosos e Sociais e o Uso de Plantas Psicoativas, Seminário Internacional: O Uso e o Abuso de Drogas (Programa Oficial)*, Salvador, Cetad/UFBa. [Results of six and a half years of observation of members of the UDV as a participant in the group]

CARIOCA, J. da S. (1999). "Santo Daime: a filosofia do século," Rio Branco, Centro de Iluminação Cristã Luz Universal (Ciclu). (Manuscript)

CARLINI, A. (2002). "Madrigal psychophármacon: recriação de proposta científico-musical para o incremento de atividades transdisciplinares na

UFRGS," self-published online at: http://www.alvarocarlini.pop.com.br.

CARLINI, A. (2005). "Histórico do madrigal psychophármacon (1988-1991)," in: *Anais do 10 Simpósio Internacional de Cognição e Artes Musicais (Sincam)*, pp. 324-333.

CARNEIRO, H. (2004). Review of: O uso ritual da ayahusca. *Rever-Revista de Estudos da Religião*, available at: http://www.pucsp.br/rever/resenha/carneiro01.htm and http://www.neip.info/downloads/l_bia1_hencar.htm.

CARVALHO, J.J. de (1999). "Um espaço público encantado: pluralidade religiosa e modernidade no Brasil," *Série Antropológica*, no. 249.

CARVALHO, J.J. de (2000). "A religião como sistema simbólico: uma atualização teórica," *Série Antropológica*, no. 285.

CARVALHO, T.B. (2005). "Em busca do encontro: a demanda numinosa no contexto religioso da União do Vegetal," Mestrado em Psicologia, PUC-RJ. [Results of research with fifteen members of the União do Vegetal (Vargem Pequena, a city in Rio de Janeiro state), 2004 to 2005], available at: http://www.Neip.info/downloads/tatiana_barbosa/tese_tati.pdf.

CASTILLA, A. (1995). *Santo Daime: fanatismo e lavagem cerebral*. Rio de Janeiro, Editora Imago.

CAZENAVE, S.O.S. (2000). "Banisteriopsis caapi e ação alucinógena," *Revista de Psiquiatria clínica*, no. 27, vol. 1. [Overview of ayahuasca with testimony from UDV members, date and place of data collection not given]

CEBUDV (1989). *Hoasca: fundamentos e objetivos*. Brasília, Sede Geral.

CEBUDV (2004). *Consolidação das leis do centro espírita beneficente União do Vegetal*. Brasília, Sede Geral, 4a ed. revisada [1a ed., 1987; 2a ed. 1991; 3a ed. 1994]

CEFLURIS (1997). *Santo Daime: normas de ritual*. Cefluris.

CEMIN, A.B. (1995a). "A cristandade entre nós: notas sobre o 'estilo de vida' daimista," in: Cetad/UFBa (eds.). I Encontro de Estudos sobre Rituais Religiosos e Sociais e o Uso de Plantas Psicoativas, Seminário Internacional: O Uso e o Abuso de Drogas (Programa Oficial), Salvador, Cetad/UFBa.

CEMIN, A.B. (1995b). "Santo Daime: conhecimento e sociabilidade," VII Encontro de Ciências Sociais ABA Norte/Nordeste, João Pessoa.

CEMIN, A.B. (1998a). "Ordem, xamanismo e dádiva: o poder do Santo Daime," Doutorado em Antropologia Social, USP.

CEMIN, A.B. (1998b). "O livro sagrado do Santo Daime: aspectos sócio-antropológicos," 21☐ Reunião da Associação Brasileira de Antropologia e da 1☐ Reunião Internacional da Teoria Arqueológica na América do Sul, Vitória, p. 68.

CEMIN, A.B. (1999). "O livro sagrado do Santo Daime," *Caderno de Criação*, no. 17, vol. 4, pp. 6-14.

CEMIN, A.B. (2001). "O poder do Santo Daime: ordem, xamanismo e dádiva," São Paulo, Terceira Margem.

CEMIN, A.B. (2004). "Os rituais do Santo Daime: 'Sistemas de Montagens Simbólicas'," in: LABATE, B.C.; ARAÚJO, W.S. (eds.). *O uso ritual da Ayahuasca*. 2nd ed. Campinas, Mercado de Letras, pp. 347-382.

CHAPUIS, F. (2002). "Arquitetura enquanto percurso e a natureza na metrópole: estudo de parque no Jaraguá," Doutorado em Arquitetura e Urbanismo, USP.

CHAVES, L.R. (2003). "A mulher urbana no Santo Daime: entre o modelo arcaico e o moderno de feminino," Mestrado em Psicossociologia de Comunidade e Ecologia Social, UFRJ. [Results of research with five female members of Santo Daime (Rio de Janeiro), 2000 to 2002]

COSTA, M.C.M.; FIGUEIREDO, M.C.; CAZENAVE, S.O.S. (2005). "Ayahuasca: uma abordagem toxicológica do uso ritualístico," *Revista de Psiquiatria clínica*, no. 32, vol. 6. [Overview of ayahuasca with testimony from users (members of the União do Vegetal and of Santo Daime). Place and date of data collected are not indicated]

COUTO, F.R (1989). "Santos e xamãs," Mestrado em Antropologia, UNB.

COUTO, F.R. (2004). "Santo Daime: rito da ordem," in: LABATE, B.C.; ARAÚJO, W.S. (eds.). *O uso ritual da Ayahuasca.* 2nd ed. Campinas, Mercado de Letras, pp. 385-410.

CUNHA, G.C.B. (1986). "O império do beija-flor," Monografia apresentada no Concurso Silvio Romero de monografias sobre Folclore/Funarte.

DIAS JUNIOR, W. (1992). "O império de Juramidam nas batalhas do astral: uma cartografia do imaginario no culto ao Santo Daime," Mestrado em Ciências Sociais, PUC-SP.

DIAS JUNIOR, W. (2004). "Diário de viagem," in: LABATE, B.C.; ARAÚJO, W.S. (eds.). *O uso ritual da Ayahuasca.* 2nd ed., Campinas, Mercado de Letras, pp. 445-471.

DOERING-SILVEIRA, E. (2003). "Avaliação neuropsicológica de adolescentes que consomem chá de Ayahuasca em contexto ritual religioso," Mestrado em Psiquiatria e Psicologia Médica, Unifesp/EPM. [Results of research with forty adolescents from the União do Vegetal in three Brazilian cities (São Paulo, Campinas, and Brasília), 2001]

DOERING-SILVEIRA, E.; DA SILVEIRA, D.X.; GROB, C.S.; LOPEZ, E.; DOBKIN DE RIOS, M. (2004). "Uso de Ayahuasca por adolescentes em contexto ritual religioso," Anais da I Conferência de Redução de Danos da América Latina e do Caribe, São Paulo.

FERNANDES, M.L.F. (2007). "Modelo alométrico para estimar biomassa de Banisteriopsis caapi (jagube, mariri) no estado do Amazonas," Mestrado em Ciência das Florestas Tropicais, Instituto Nacional de Pesquisas da Amazônia. (Project in development)

FERNANDES, V.F. (1986). *História do povo Juramidam: introdução à cultura do Santo Daime.* Manaus, Suframa.

FERNANDES, V.F. (ed.). *Santo Daime cultura amazônica.* 2nd ed., São Paulo, Joruês.

FERREIRA, C.A. (2006). "O vinho das almas: xamanismo, sincretismo e cura na doutrina do Santo Daime," Mestrado em Ciências da Religião, PUC-SP, available at: http://alto-das-estrelas.blogspot.com/2007/03/pesquisa-em-andamento-discute-se-o.html. (Project in development)

FIGUEIREDO, M. (1992). "A experiência no Santo Daime," in: COSTA, M.S.R. (ed.). *Karma ou destino?: trajetória de uma mãe-de-santo.* Rio de Janeiro, Salamandra, pp. 153-162.

FREITAS, L.C.T. de (2006a). *A rainha da floresta: A missão daimista de evangelização.* Online book. Juramidam, available at: http://www.juramidam.jor.br/rainha/index.html.

FREITAS, L.C.T. de (2006b). O mensageiro: o replantio daimista da doutrina cristã. Online book. *Juramidam,* available at: http://www.juramidam.jor.br/mensageiro/index.html.

GABRICH, D. de C.P. (2005). "O trabalho oculto e exotérico de Raimundo Irineu Serra. Primeiro Encontro Brasileiro de Xamanismo," *Núcleo de Estudos Interdisciplinares sobre Psicoativos – Neip,* available at: www.neip.info/downloads/debora/oculto_exoterico.pdf.

GABRICH, D. de C.P. (2006). "Ecletismo religioso: um estudo de caso das origens amazônicas da doutrina do Santo Daime e sua mutação na metrópole de Belo Horizonte, Minas Gerais," *CD ROM e Libro de Memorias. Foro Latinoamericano Memória e Identidad.* Montevideo, Signo.

GENTIL, L.R.B.; GENTIL, H.S. (2004). "O uso de psicoativos em um contexto religioso: a União do Vegetal," in: LABATE, B.C.; ARAÚJO, W.S. (eds.). *O uso ritual da Ayahuasca.* 2nd ed., Campinas, Mercado de Letras, pp. 559-569.

GODOY, A.S. de M. (2006). "A Suprema Corte Norte-Americana e o julgamento do uso de Huasca pelo Centro Espírita Beneficente União do Vegetal (UDV): colisão de princípios: liberdade religiosa v. repressão a substâncias alucinógenas: um estudo de caso." *Revista Jurídica,* no. 8, vol. 79, available at: http://www.planalto.gov.br/ccivil_03/revista/Rev_79/artigos/Arnaldo_rev79.htm.

GOMES, G. (1998). *Alucinógenos: informativos farmacológicos e correlatos à matéria do DMT ou dimetiltriptamina: (princípio ativo) do Daime ou "chá do Santo Daime."* São Paulo, Editora Oliveria Mendes.

GOULART, S.L. (1995). "Modernidade e psicoativos no culto do Santo Daime," in: Cetad/UFBa (eds.). *I Encontro de Estudos sobre Rituais Religiosos e Sociais e o Uso de Plantas Psicoativas, Seminário Internacional: O Uso e o Abuso de Drogas (Programa Oficial)*, Salvador, Cetad/UFBa.

GOULART, S.L. (1996a). "A história do encontro do mestre Irineu com a Ayahuasca: mitos fundadores da religião do Santo Daime," *Núcleo de Estudos Interdisciplinares sobre Psicoativos – Neip*, available at: http://www.neip.info/downloads/t_sandra_encontro.pdf.

GOULART, S.L. (1996b). "Raízes culturais do Santo Daime," Mestrado em Antropologia Social, USP.

GOULART, S.L. (1999). "Contrastes e continuidades entre os grupos do Santo Daime e da Barquinha," *Núcleo de Estudos Interdisciplinares sobre Psicoativos – Neip*, 1999, available at: http://www.Neip.info/downloads/TEXTO%20Sandra%20Goulart%20-20JORNADASRELIGIOSAS.pdf.

GOULART, S.L. (2003). "A construção de fronteiras religiosas através do consumo de um psicoativo: as religiões da Ayahuasca e o tema das drogas," *Núcleo de Estudos Interdisciplinares sobre Psicoativos – Neip*, available at: http://www.neip.info/downloads/t_san2.pdf.

GOULART, S.L. (2004a). "Contrastes e continuidades em uma tradição Amazônica: as religiões da Ayahuasca," Doutorado em Ciências Sociais, Unicamp.

GOULART, S.L. (2004b). "O contexto de surgimento do culto do Santo Daime: formação da comunidade e do calendário ritual," in: LABATE, B.C.; ARAÚJO, W.S. (eds.). *O uso ritual da Ayahuasca*. 2nd ed. Campinas, Mercado de Letras, pp. 277-301.

GOULART, S.L. (2005). "Contrastes e continuidades em uma tradição religiosa amazônica. Os casos do Santo Daime, da Barquinha e da UDV," in: LABATE,

B.C.; Goulart, S.L. (eds.). *O uso ritual das plantas de poder.* Campinas, Mercado de Letras, pp. 355-396.

GOULART, S.L. (2007a). "As religiões da Ayahuasca e o tema das drogas: história e construção de identidade," in: Coggiola, O. (ed.). *Caminhos da história.* São Paulo, Xamã Editora, pp. 79-87.

GOULART, S.L. (2007b). "Estigmas de cultos Ayahuasqueiros," XIV Jornadas Sobre Alternativas Religiosas en América Latina. Buenos Aires, Unsam.

GREGORIM, G. (1991). *Santo Daime: estudos sobre simbolismo, doutrina e povo de Juramidam.* São Paulo, Ícone Editora.

GROB, C.S.; McKENNA, D.J.; CALLAWAY, J.C.; BRITO, G.S.; NEVES, E.S.; OBERLENDER, G.; SAIDE, O.L.; LABIGALINI, E.; TACLA, C.; MIRANDA, C.T.; STRASSMAN, R.J.; BOONE, K.B. (1996). "Farmacologia humana da hoasca, planta alucinógena usada em contexto ritual no Brasil: I. Efeitos psicológicos," *Informação Psiquiátrica,* no. 15, vol. 2, pp. 39-45. [Results of the Hoasca Project], available at: http://www.udv.org.br/portugues/downloads/01.rtf.

GROB, C.S.; McKENNA, D.J.; CALLAWAY, J.C.; BRITO, G.S.; ANDRADE, E.O.; OBERLENDER, G.; SAIDE, O.L.; LABIGALINI, E.; TACLA, C.; MIRANDA, C.T.; STRASSMAN, R.J.; BOONE, K.B.; NEVES, E.S. (2004). "Farmacologia humana da hoasca, planta alucinógena usada em contexto ritual no Brasil: efeitos psicológicos," in: LABATE, B.C.; ARAÚJO, W.S. (eds.). *O uso ritual da Ayahuasca.* 2nd ed., Campinas, Mercado de Letras, pp. 653-669. [Results of the Hoasca Project; Portuguese translation of Grob et. al. (1996)]

GROISMAN, A. (1990). "Santo Daime: notas sobre a luz xamânica da rainha da floresta," Reunião da Associação Brasileira de Antropologia, Florianópolis.

GROISMAN, A. (1991). "'Eu venho da Floresta': Ecletismo e práxis xamânica daimista no 'Céu do Mapiá'," Mestrado em Antropologia Social, UFSC.

GROISMAN, A. (1993). "A limpeza da corrente: ritual e significado na cura daimista," IX Reunião Regional da Associação Brasileira de Antropologia,

Florianópolis.

GROISMAN, A. (1994). "Messias, milênio e salvação: motivação e engajamento na doutrina do Santo Daime," in: BARABAS, A. (ed.). *Religiosidad y resistencia indígenas hacia el fin del milenio,* no. 11, pp. 269-286.

GROISMAN, A. (1994). "Santo Daime: uma religião substancial," Congresso Internacional As Novas Religiões: A Expansão Internacional dos Movimentos Religiosos e Mágicos, Recife.

GROISMAN, A. (1995). "Santo Daime: notas sobre a luz xamânica da rainha da floresta," in: LANGDON, E.J. (ed.). *Xamansimo no Brasil. Novas perspectivas.* Florianópolis, EdUFSC, pp. 333-352.

GROISMAN, A. (1996). "O Santo Daime e a nova peregrinação," *Estudos leopoldenses, série Ciências Humanas,* no. 32, vol. 146, pp. 111-119.

GROISMAN, A. (1995). "'Eu sou o guia que vai atrás': as vozes da cura espiritual no Santo Daime," in: Cetad/UFBa (eds.). *I Encontro de Estudos sobre Rituais Religiosos e Sociais e o Uso de Plantas Psicoativas, Seminário Internacional: O Uso e o Abuso de Drogas (Programa Oficial),* Salvador, Cetad/UFBa.

GROISMAN, A. (1999). "Eu venho da floresta. Um estudo sobre o contexto simbólico do uso do Santo Daime," Florianópolis, Editora da UFSC.

GROISMAN, A. (2002). "O lúdico e o cósmico: rito e pensamento entre daimistas holandeses," *Antropologia em Primeira Mão,* no. 53, pp. 1-20.

GROISMAN, A. (2004a). "Trajetos, fronteiras e reparações," *Antropologia em Primeira Mão,* no. 73, pp. 1-26, available at: http://www.antropologia.ufsc.br/73.%20alberto-trajetos.pdf.

GROISMAN, A. (2004b). "Missão e projeto: motivos e contingências nas trajetórias dos agrupamentos do Santo Daime na Holanda," *Rever: Revista de Estudos da Religião,* no. 4, vol. 1, pp. 1-18.

GROISMAN, A. (2004c). Review of: *O uso ritual da Ayahuasca. Ilha,* no. 6, vols. 1-2, pp. 235-240, available at: http://www.neip.info/downloads/resenha_uso_aya_ilha.pdf.

GROISMAN, A. (2007). "Argumentos jurídicos e fundamentos etnográficos: religião e saúde como categorias de negociação, no contexto do debate sobre a legalização das Religiões Ayahuasqueiras Brasileiras nos EUA," Projeto de Pós-Doutorado, Arizona State University.

GUIMARÃES, M.B.L. (1992). "A 'lua branca' de Seu Tupinambá e de mestre Irineu: estudo de caso de um terreiro de umbanda," Mestrado em Ciências Sociais, UFRJ.

GUIMARÃES, M.B.L. (1996). "Umbanda e Santo Daime na Lua Branca de Lumiar: estudo de caso de um terreiro de umbanda," *Religião & Sociedade*, no.17, vol. 1-2, pp. 124-139.

GUIMARÃES, M.B.L. (2005). "A Lua Branca de Seu Tupinambá e de mestre Irineu: estudo de caso de um terreiro de umbanda," XIII Jornadas sobre Alternativas Religiosas na América Latina, Porto Alegre.

HUMANUS. Campinas, Sama Editora. 2000 a 2005. [Magazine with publications about the Centro Espiritual Beneficente União do Vegetal, led by Joaquim José de Andrade Neto and based in Campinas (São Paulo state)]

JACCOUD, S. (1992). *O terceiro testamento: um fato para a história*. Goiânia, Página Um.

JORNAL ALTO FALANTE. Newsletter published by the Centro Espírita Beneficente União do Vegetal (CEBUDV) in print and online, 1988 to present. (Access restricted to members)

LABATE, B.C. (2000). "A reinvenção do uso da Ayahuasca nos centros urbanos," Mestrado em Antropologia Social, Unicamp.

LABATE, B.C. (2004a). "A literatura brasileira sobre as religiões ayahuasqueiras," in: LABATE, B.C.; ARAÚJO, W.S. (eds.). *O uso ritual da Ayahuasca*. 2nd ed., Campinas, Mercado de Letras, pp. 231-273.

LABATE, B.C. (2004b). *A reinvenção do uso da Ayahuasca nos centros urbanos*. Campinas, Mercado de Letras.

LABATE, B.C. (2004c). *Ayahuasca mamancuna merci beaucoup: diversificação e internacionalização do vegetalismo ayahuasqueiro peruano.* Campinas. (Manuscript)

LABATE, B.C. (2005). "Dimensões legais, éticas e políticas da expansão do consumo da Ayahuasca," in: LABATE, B.C.; GOULART, S.L. (eds.). *O Uso ritual das plantas de poder.* Campinas, Mercado de Letras, pp. 397-457.

LABATE, B.C.; ARAÚJO, W.S. (eds.). (2002). *O uso ritual da Ayahuasca.* 1st ed., Campinas, Mercado de Letras.

LABATE, B.C.; ARAÚJO, W.S. (eds.). (2004). *O uso ritual da Ayahuasca.* 2a ed., revista e ampliada, Campinas, Mercado de Letras.

LABATE, B.C.; Pacheco, G. (2004). "Matrizes maranhenses do Santo Daime," in: LABATE, B.C.; ARAÚJO, W.S. (eds.). *O uso ritual da Ayahuasca.* 2nd ed., Campinas, Mercado de Letras, pp. 303-344.

LABATE, B.C.; Goulart, S.L. (eds.). (2005a). *O uso ritual das plantas de poder.* Campinas, Mercado de Letras.

LABATE, B.C.; Goulart, S.L. (2005b). "Religiões ayahuasqueiras," IHU On-Line, Revista do Instituto Humanitas Unisinos, no. 169, pp. 90-94, available at: http://www.unisinos.br/ihu_online/uploads/edicoes/1158349107.48pdf.pdf.

LABIGALINI JR., E. (1988). "O uso de Ayahuasca em um contexto religioso por ex-dependentes de álcool: um estudo qualitativo," Mestrado em Saúde Mental, Unifesp/EPM. [Results of research with four (ex-) alcohol-dependent members of the União do Vegetal. Araçariguama (SP), 1998], available at: http://www.udv.org.br/portugues/downloads/13.rtf.

LABIGALINI JR., E.; MIRANDA, C.T.; TACLA, C. (1995). "O uso ritualizado de Ayahuasca: alternativa terapêutica para alcoolismo," in: Cetad/UFBa (eds.). I Encontro de Estudos sobre Rituais Religiosos e Sociais e o Uso de Plantas Psicoativas, Seminário Internacional: O Uso e o Abuso de Drogas (Programa Oficial), Salvador, Cetad/UFBa. [Results of research with (ex-) alcohol-dependent members of the União do Vegetal and results of the Hoasca Project]

LIMA, F.A. de S. (2002). "O poder do Santo Daime, ou, com quantos paus se faz uma canoa," Review of: *Ordem, xamanismo e dádiva: o poder do Santo Daime, Labirinto, Revista Eletrônica do Centro de Estudos do Imaginário*, no. 4, available at: http://www.cei.unir.br/res2.html.

LIMA, F.A. de S. (2004). Review of: *Ordem, xamanismo e dádiva: o poder do Santo Daime, Revista de Estudos da Religião – Rever,* available at: http://www.pucsp.br/rever/resenha/cemin01.htm.

LIMA, F.A. de S.; NAVES, M.B.; MOTTA, J.M.C.; DI MIGUELI, J.C.V.; BRITO, G.S. et al. (1998). "Sistema de Notificação e Monitoramento Psiquiátrico em Instituição de Usuários do Chá Hoasca – União da Vegetal," XVI Congresso Brasileiro de Psiquiatria, São Paulo, available at: http://www.udv.org.br/portugues/downloads/12.rtf. [Report on monitoring of psychiatric crises in centers of the União do Vegetal (states of São Paulo, Paraná, Santa Catarina, and Rio Grande do Sul) from December 1995 to May 1998]

LIMA, F.A. DE S.; NAVES, M.B.; MOTTA, J.M.C.; DI MIGUELI, J.C.V.; BRITO, G.S. et al. (2002). "Sistema de Monitoramento Psiquiátrico de Usuários do Chá Hoasca," *Revista Brasileira de Psiquiatria,* no. 24, (supl. 2). [Report on monitoring of psychiatric crises in centers of the União do Vegetal (states of São Paulo, Paraná, Santa Catarina, and Rio Grande do Sul) from 1996 to 2000]

LIMA, S.A.S. (2005). "Experiências e símbolos de transformação na doutrina da floresta," Mestrado em Ciências da Religião, UFJF, available at: http://alto-das-estrelas.blogspot.com/2005/10/dissertao-de-mestrado-em-cincia-da.html.

LODI, E. (2004). *Estrela da minha vida: histórias do sertão caboclo.* Brasília, Edições Entre Folhas, available at: http://alto-das-estrelas.blogspot.com/2005/11/lanamento-do-livro-estrela-da-minha_22.html and http://alto-das-estrelas.blogspot.com/2005/07/lanado-livro-sobre-unio-do-vegetal.html.

LUNA, L.E. (1995a). "Ayahuasca em cultos urbanos brasileiros. Estudo contrastivo de alguns aspectos do Centro Espírita e Obra de Caridade Príncipe Espadarte Reino da Paz (a Barquinha) e o Centro Espírita Beneficente União do Vegetal," Trabalho apresentado para o concurso de professor adjunto em antropologia. Departamento de Ciências Sociais da UFSC.

LUNA, L.E. (1995b). "A Barquinha, uma nova religião do Rio Branco, na Amazônia brasileira," in: Cetad/UFBa (eds.). *I Encontro de Estudos sobre Rituais Religiosos e Sociais e o Uso de Plantas Psicoativas, Seminário Internacional: O Uso e o Abuso de Drogas (Programa Oficial)*. Salvador, Cetad/UFBa.

LUZ, C.R.M da (1999). "A internet na União do Vegetal: reflexos da comunicação globalizada na estrutura de uma religião iniciática com tradição de transmissão oral de conhecimento," Mestrado em Comunicação Social, UFRJ.

MacRAE, E. (1989). "Observações sobre o documento do grupo de trabalho do Conselho Federal de Entorpecentes," in: FERNANDES, V.F. (ed.). *Santo Daime cultura amazônica*. 2nd ed., São Paulo, Joruês.

MacRAE, E. (1992). *Guiado pela lua. Xamanismo e uso ritual da Ayahuasca no culto do Santo Daime*. São Paulo, Editora Brasiliense.

MacRAE, E. (1994). "A importância dos fatores socioculturais na determinação da política oficial sobre o uso ritual da Ayahuasca," in: ZALUAR, A. (ed.). *Drogas e cidadania*. São Paulo, Editora Brasiliense, pp. 31-45.

MacRAE, E. (1997a). "O Santo Daime e outras religiões brasileiras," V Encontro de Antropólogos do Norte e Nordeste, Antropologia Cultural: memória, tradição e perspectivas, Recife.

MacRAE, E. (1997b). "Possessão e vôo xamânico no Santo Daime," V Congresso Afro-Brasileiro, Salvador.

MacRAE, E. (1998). "A utilização de substâncias psicoativas na religião do Santo Daime," VII Jornadas sobre alternativas religiosas na América Latina, São Paulo.

MacRAE, E. (1999). "O uso da Ayahuasca nos rituais de cura do Santo Daime," in: Greiner, C.; Bião, A. (eds.). *Etnocenologia: textos selecionados*. São Paulo, Annablume, pp. 109-118.

MacRAE, E. (2000). "O Ritual do Santo Daime como espetáculo e performance," in: TEIXEIRA, J.G.; GUSMÃO, R. (eds.). *Performance, cultura e espetacularidade,* no. 2. Brasília, Editora Transe, pp. 75-84.

MacRAE, E. (2004). "Um pleito pela tolerância entre as diferentes linhas ayahuasqueiras," in: LABATE, B.C.; ARAÚJO, W.S. (eds.). *O uso ritual da Ayahuasca.* 2nd ed., Campinas, Mercado de Letras, pp. 493-505.

MacRAE, E. (2005). "Santo Daime e Santa Maria: usos religiosos de substâncias psicoativas lícitas e ilícitas," in: LABATE, B.C.; GOULART, S.L. (eds.). *O uso ritual das plantas de poder.* Campinas, Mercado de Letras, pp. 459-485.

MacRAE, E. (2007). "A elaboração das políticas públicas brasileiras em relação ao uso religioso da Ayahuasca," *XIV Jornadas Sobre Alternativas Religiosas en América Latina,* Buenos Aires, Uusam.

MAIA NETO, F.J. (2003). *Contos da Lua Branca: histórias do mestre Raimundo Irineu Serra e de sua ora espiritual, contadas por seus contemporâneos.* Rio Branco, Fundação Elias Mansour.

MARGARIDO, F.L.; ARAÚJO NETO, F.H. de (eds.). (2005). *Centro espírita e culto de oração "Casa de Jesus – Fonte de Luz." Mestre Daniel. História com a Ayahuasca.* Rio Branco, Fundação Garibaldi Brasil.

MELO, A.G. de (2007a). *Viagens ao Juruá (Journeys to the Juruá).* Rio de Janeiro, Edição Cefluris, available at: http://alto-das-estrelas.blogspot.com/2007/10/padrinho-alfredo-gregrio-de-melo.html.

MELO, A.G. de (2007b). *Padrinho Sebastião. Biografia Versejada.* Caxias do Sul, Edição Ecovila rainha da floresta, available at: http://alto-das-estrelas.blogspot.com/2007/10/padrinho-alfredo-gregrio-de-melo.html.

MELO, R.V. (2006). "Beber na fonte: adesão e a transformação do self na União do Vegetal" (provisional title), Doutorado em Antropologia Social, UnB. (Project in development)

MELLO, P.B.B. (1995). "O Santo Daime como caminho de individuação," in: Cetad/UFBa (eds.). *I Encontro de Estudos sobre Rituais Religiosos e Sociais e o*

Uso de Plantas Psicoativas, Seminário Internacional: O Uso e o Abuso de Drogas (Programa Oficial). Salvador, Cetad/UFBa.

MENEZES, M.R. (2005). "Hospitalidade e dádiva: o caso do Santo Daime," Mestrado em Hospitalidade, Universidade Anhembi Morumbi.

MERCANTE, S.M. (2000). "Barquinha, caridade e cura," 22nd Reunião Brasileira de Antropologia, Brasilia.

MERCANTE, S.M. (2002). "Ecletismo, caridade e cura na Barquinha da Madrinha Chica," *Humanitas*, no. 18, vol. 2, pp. 47-60, available at: http://www.neip.info/downloads/barquinha.pdf.

MERCANTE, S.M. (2003a). Review of: O uso ritual da Ayahuasca, *Horizontes Antropológicos*, no. 19, pp. 323-330, available at: http://www.neip.info/downloads/l_bia1_mercadante.pdf.

MERCANTE, S.M. (2003b). Review of: O uso ritual da Ayahuasca. *Campos*, no. 4, pp. 211-217, available at: http://www.neip.info/downloads/l_bia1_resenha_campos.htm.

MERCANTE, S.M. (2004). "O papel das visões experienciadas durante rituais com ingestão de Ayahuasca na consciência durante os processos de doença e cura," 24th Reunião Brasileira de Antropologia. Olinda.

MERCANTE, S.M. (2005). Review of: A reivenção do uso da Ayahuasca nos centros urbanos. *Campos*, no. 6, vols. 1-2, pp. 229-231, available at: http://www.neip.info/downloads/resenha_reiv_mercante.pdf.

MERCANTE, S.M. (2007). "Estudo do uso terapêutico de Ayahuasca entre moradores de rua na cidade de São Paulo," Projeto de pesquisa de Pós-Doutorado Junior, UFSC. (Project in development)

METZNER, R. (ed.) (2002). *Ayahuasca: alucinógenos, consciência e o Espírito da Natureza*. Rio de Janeiro, Gryphus.

MIKOSZ, J.E. (2005). Arte e Ayahuasca: Imagens e Simbologia das Espirais. Doutorado em Interdisciplinar em Ciências Humanas, Universidade Federal de Santa Catarina (project in development), available at: http://alto-das-estrelas.blogspot.com/2007/10/pesquisa-em-andamento-sobre-relao-entre.html.

MIKOSZ, J.E. (2006a). "Substâncias psicoativas e religião," *Cadernos de pesquisa do doutorado interdisciplinar em Ciências Humanas* (DICH/UFSC), no. 79, available at: http://www.cfh.ufsc.br/ dich/TextoCaderno79.pdf.

MIKOSZ, J.E. (2006b). "O vórtice entóptico: estados não ordinários de consciência e fenômenos visuais," *Anais do V Forum de Pesquisa Científica em Arte da Embap* (Escola de Música e Belas Artes do Paraná), available at: http://www.anais.embap.br/forum20062007/textuais/09%20 Jose%20Eliezer%20 Mikosz.pdf.

MILANEZ, W. (2001). *Oaska: o Evangelho da Rosa*. 2nd ed., Campinas, Sama Editora. [1st ed. 1988]

MILANEZ, W. (2003). *Oaska: o misterioso chá da Amazônia – Relatos de experiências*. 3rd ed., Campinas, Sama Editora.

MONTEIRO, C. S. (1982a). "O Sistema de Juramidã: religião popular no Acre no seringal Céu do Mapiá," XI Reunião da Associação Brasileira de Antropologia, São Paulo.

MONTEIRO, C. S. (1982b). *O povo de Juramidã no seringal Rio do Ouro*. (Manuscript)

MONTEIRO, C. S. (1983a). "O Palácio Juramidam: Santo Daime: um ritual de transcendência e despoluição," Mestrado em Antropologia Cultural, UFPE, available at: http://www.neip.info/ downloads/clodomir/teseClodomir.pdf.

MONTEIRO, C. S. (1983b). "Religiosidade popular no contexto social e econômica da Amazônia," IV Encontro de Pesquisadores da Amazônia, Rondônia.

MONTEIRO, C. S. (1985a). "Ritual de tratamento e cura," I Simpósio de Saúde Mental da Amazônia. Santarém.

MONTEIRO, C. S. (1985b). "Culto do Santo Daime, possíveis sobrevivências culturais africanas," Seminário Identidade e Resistência Cultural do Negro

Brasileiro. São Luís.

MONTEIRO, C. S. (1985c). "A questão da realidade na Amazônia," IV Encontro Inter-Regional de Cientistas Sociais do Brasil. A Amazônia em questão, Manaus, pp. 89-141.

MONTEIRO, C. S. (1986). "Xamanismo, cura e organização social," V Encontro de Pesquisadores da Amazônia. Manaus.

MONTEIRO, C. S. (1989). "Ritual de tratamento e cura: contexto religioso, social" (O culto do Santo Daime no Acre), Simpósio Representações e práticas das Medicinas Tradicionais (indígena, cabocla, etc.), Belém.

MONTEIRO, C. S. (1992). "O uso cultural de psicoativos," Paper presented to the Symposium on medicinal plants of Amazônia at the Earth Summit (ECO – 92), Rio de Janeiro.

MONTEIRO, C. S. (2004). "O uso ritual da Ayahuasca e o reencontro de duas tradições. A Miração e a Incorporação no Culto do Santo Daime," in: LABATE, B.C.; ARAÚJO, W.S. (eds.). *O uso ritual da Ayahuasca*. 2nd ed., Campinas, Mercado de Letras, pp. 413-443.

MORAIS, A.F. (2005). "O ethos e o futuro na Vila Céu do Mapiá, Amazonas," Tese de Doutorado em Psicologia Social, USP.

MOREIRA, P. (2006). "Eu venho de longe: uma história de vida de Raimundo Irineu Serra, fundador do Santo Daime," Mestrado em Ciências Sociais, UFBa. (Project in development)

MORTIMER, L. (2000). *Bença Padrinho*. São Paulo, Céu de Maria, São Paulo.

MORTIMER, L. (2001). *Nosso Senhor Aparecido na Floresta*. Céu de Maria, São Paulo.

MOURÃO, J. (1995). *Tragédia na Seita do Daime*. Rio de Janeiro, Editora Imago.

NAKAMAKI, H. (1994). "Quem não toma o chá não tem alucinações: epidemiologia de religiões alucinógenas no Brasil," in: PELLEGRINI, A.; NA-

KAMAKI, H. (eds.). *Possessão e procissão: religiosidade popular no Brasil. Senri Ethnological Reports* no. 1. National Museum of Ethnology, p. 61-86.

NASCIMENTO, S.B. do (2005). *No brilho da lua branca.* Rio Branco, Fundação Garibaldi Brasil.

NOGUEIRA, S. (1994). "Núcleo urbano florestal: modelo arquetípico da cultura e do homem amazônico," Doutorado em Arquitetura e Urbanismo USP.

NOGUEIRA, S. (1996). "Núcleo urbano florestal," Congresso Internacional de Arquitetura. Barcelona.

OLIVEIRA, I.L. (2007). "Santo Daime: um sacramento vivo, uma religião em formação," Doutorado em História, UnB.

OLIVEIRA, J.E.B. de (2004). "Santo Daime: uma religião brasileira: estudo de sua trajetória através dos hinos," Doutorado em Sociologia, UFC. (Project in development)

OLIVEIRA, J.E.B. de (2006). *Raimundo Irineu Serra: mestre Império Juramidam,* Fortaleza, Tupynamquim Editora, available at: http://alto-das-estrelas.blogspot.com/2007/05/cordel-sobre-vida-do-mestre-irineu.html.

OLIVEIRA, M. (1993). Minha viagem ao centro do Daime. São Paulo, Editora Saraiva.

OLIVEIRA, R.M de (2002). "De folha e cipó é a capelinha de São Francisco: a religiosidade popular na cidade de Rio Branco: Acre (1945-1958)," Mestrado em História, UFPE.

PACHECO, G.; LABATE, B.C. (2007). "Hinos e chamadas, abrindo as portas do céu," *XIV Jornadas Sobre Alternativas Religiosas en América Latina,* Buenos Aires, Unsam.

PACHECO, G. (1999). "Os hinos são as correntes: notas para um estudo antropológico da música no Santo Daime," PPGAS/Museu Nacional, Rio de Janeiro. (Manuscript)

PACHECO, G. (2000). "Recebendo um hino: inspiração e composição na música daimista," Encontro Internacional de Etnomusicologia: Música Africanas e Indígenas em 500 anos de Brasil, Belo Horizonte.

PASKOALI, V.P. (1998). "A Barquinha de Antônio Geraldo: história, morte e simbolismo," VI Seminário de Bolsistas de Iniciação Científica, Rio Branco.

PASKOALI, V.P. (2002). "A cura enquanto processo identitário na Barquinha: o sagrado no cotidiano," Mestrado em Ciências Sociais, PUC-SP.

PELÁEZ, M.C. (1994). "No mundo se cura tudo. Interpretações sobre a `Cura Espiritual' no Santo Daime," Mestrado em Antropologia Social, UFSC.

PELÁEZ, M.C. (2004). "Santo Daime, transcendência e cura. Interpretações sobre as posssibilidades terapêuticas da bebida ritual," in: LABATE, B.C.; ARAÚJO, W.S. (eds.). *O uso ritual da Ayahuasca*. 2nd ed., Campinas, Mercado de Letras, pp. 473-491.

PEREIRA, E. (2003). Review of: *O uso ritual da Ayahuasca, Revista Brasileira de Ciências Sociais,* no. 18, vol. 52, pp. 203-207.

PEREIRA, M.T. (2003). "Arquitetura como um microcosmo: religiosidade e representação do espaço na Comunidade do Matutu-MG," Mestrado em Extensão Rural, UFV, available at: http://alto-das-estrelas.blogspot.com/2005/12/tese-de-arquitetura-sobre-comunidade.html.

PEREIRA, M.T.; DOULA, S.M. (2002). "Arquitetura como um microcosmo: religiosidade e representação do espaço na comunidade do Matutu," XII Simpósio de Iniciação Científica e II Mostra da Pós-Graduação, Viçosa.

PEREIRA, M.T.; DOULA, S.M. (2003). "Arquitetura como um microcosmo: uma análise arquitetônica e antropológica do espaço construído na comunidade do Matutu," I Simpósio de Arquitetura e Conceito, Belo Horizonte.

PEREIRA, N. (1979). *A Casa das Minas: contribuição ao estudo das sobrevivências do culto dos voduns do panteão daomeano no Estado do Maranhão*. Petrópolis, Vozes. [1947]

PEREIRA, O.C. de C. (2007). "Análise do artigo 'Banisteriopsis caapi: ação alucinógena e uso ritual'," *Núcleo de Estudos Interdisciplinares sobre Psicoativos*

– *Neip,* available at: http://www.neip. info/downloads/o_pereira/tcc_cazenave.pdf.

PINTO, T. de O. (2001). "Som e música. Questões de uma antropologia sonora," *Revista de Antropologia,* no. 44, vol. 1, pp. 222-286.

REHEN, L.K. (2005). "Música e emoção: a linguagem do Santo Daime," IX ABANNE, Reunião de Antropólogos do Norte e Nordeste, Manaus.

REHEN, L.K. (2006a). "Cantando na primeira pessoa: texto e contexto dos hinos no culto do Santo Daime," III Encontro Internacional da Associação Brasileira de Etnomusicologia (Abet), São Paulo.

REHEN, L.K. (2006b). "Os hinários do Santo Daime: algumas considerações sobre tempo e música," Seminário Interno do Programa da Pós-Graduação em Ciências Sociais, Rio de Janeiro.

REHEN, L.K. (2006c). "Sentimento e espontaneidade: o caso da oferta de músicas na religião do Santo Daime," III Encontro Temático "Antropologia das Emoções": Linha de pesquisa Perspectivas da Subjetividade, Rio de Janeiro.

REHEN, L.K. (2006d). "Um Novo 'Tempo': o papel da música nos rituais do Santo Daime e a construção da realidade místico-religiosa," 25 Reunião Brasileira de Antropologia, Goiânia.

REHEN, L.K. (2007a). "Os hinos são presentes: algumas considerações sobre a oferta de cânticos no Santo Daime," *Núcleo de Estudos Interdisciplinares sobre Psicoativos – Neip,* available at: http://www.neip.info/downloads/lucas/texto%20Lucas%20Neip.pdf.

REHEN, L.K. (2007b). "Micropolítica dos sentimentos na religião do Santo Daime," VII RAM, Reunião de Antropologia do Mercosul, Porto Alegre.

REHEN, L.K. (2007c). "Recebido e ofertado: a natureza dos hinos nos rituais do Santo Daime," Mestrado em Ciências Sociais, Uerj, available at: http://alto-das-estrelas.blogspot.com/2007/ 04/defendida-dissertao-de-mestrado-em.html.

REVISTA DO CENTENÁRIO (1992). *Edição Comemorativa dos 100 anos de mestre Irineu,* Rio de Janeiro, Ed. Beija Flor.

RIBEIRO, F. (2005). *Os incas, as plantas de poder e um tribunal espanhol.* Rio de Janeiro, Editora Mauad, available at: http://alto-das-estrelas.blogspot.com/2005/12/comentrio-sobre-o-livro-os-incas-as.html and http://alto-das-estrelas.blogspot.com/ 2005/08/novidade-literria-sobre-o-processo.html.

RICCIARDI, G.S. (2007a). "O uso da Ayahuasca e a experiência de alívio, transformação e cura na União do Vegetal (UDV)," XIII Encontro de Ciências Sociais Norte e Nordeste, Maceió.

RICCIARDI, G.S. (2007b). "O uso da Ayahuasca e a experiência de alívio e cura em um contexto religioso: a União do Vegetal," Mestrado em Ciências Sociais, UFBA. (Project in development)

RODRIGUES, D. (1998). *Mistérios e encantos da Oaska: Danielle Rodrigues entrevista o mestre da União do Vegetal.* 3rd ed., Campinas, Sama Editora.

ROMERO, A.H.F. (2003). "Santo Daime, educação e cidadania," *VIII Encontro dos Antropólogos do Norte e Nordeste* – Abanne, São Luís.

ROMERO, A.H.F. (2004a). "Autopoise e educação no movimento do Santo Daime," Mestrado em Educação, Ufam.

ROMERO, A.H.F. (2004b). "Novos movimentos sociais, cidadania e educação na vila do céu do Mapiá," Amazônida. *Revista do Programa de pós-graduação da Faculdade de Educação da Universidade Federal do Amazonas,* no. 9, vol. 1, pp. 27-39.

ROSE, I.S. de (2004). "A experiência da cura espiritual no Santo Daime," XXIV Reunião Brasileira de Antropologia, Programa e Resumos XXIV RBA, Olinda.

ROSE, I.S. de (2005a). "O 'sistema de cuidados da saúde' do Céu da Mantiqueira e o continuum espiritual-terapêutico: análises sobre a 'cura' no Santo Daime," *Núcleo de Estudos Interdisciplinares sobre Psicoativos – Neip,* available at: www.neip.info/textos_colaboradores.htm.

ROSE, I.S. de (2005b). "Fronteiras entre espiritualidade e terapia: algumas reflexões sobre o uso terapêutico da Ayahuasca," VI Reunión de Antropología del Mercosur, Identidad, Fragmentación y Diversidad, Montevideo.

ROSE, I.S. de (2005c). "Espiritualidade, terapia e cura. Um estudo sobre a expressão da experiência no Santo Daime," Mestrado em Antropologia Social, UFSC.

ROSE, I.S. de (2006a). "Repensando as fronteiras entre espiritualidade e terapia: reflexões sobre a 'cura' no Santo Daime," *Campos,* no. 7, vol. 1, pp. 35-52, available at: http://www.neip.info/ mainwebsite_html/downloads/isabel/art_isabel_04.pdf.

ROSE, I.S. de (2006b). "Cura espiritual, biomedicina e intermedicalidade no Santo Daime," 25th Reunião Brasileira de Antropologia, Saberes e Práticas Antropológicas, Desafios para o século XXI. Goiânia.

ROSE, I.S. de (2007a). "Cura entre colinas verdes: trabalhos espirituais, plantas e culinária. Reflexões sobre experiências de campo numa comunidade do Santo Daime," in: Bonetti, A. de L.; Fleischer, S. (eds.). *Entre saias justas e jogos de cintura: gênero e etnografia na antropologia brasileira recente.* Florianópolis, Editora Mulheres/EDunisc, pp. 329-352, available at: http://www.neip.info/mainwebsite_html/downloads/isabel/art_isabel_02.pdf and http://alto-das-estrelas.blogspot.com/2007/04/livro-discute-bastidares-da-pesquisa-de.html.

ROSE, I.S. de (2007b). "Cura Aliança das medicinas: um (re)encontro entre os índios Guarani, o Santo Daime e o Caminho Vermelho," Doutorado em Antropologia Social, UFSC. (Project in development)

ROSE, I.S. de (2007). "Cura de longe ou de perto? Subjetividade, experiência pessoal e estados modificados de consciência nas pesquisas antropológicas," Ensaio de Qualificação de Doutorado em Antropologia Social, UFSC.

SANTOS, R.G. (2004). Review of: *O uso ritual da Ayahuasca. Labirinto, Revista Eletrônica do Centro de Estudos do Imaginário,* no. 6, available at: http://www.cei.unir.br/res41.html and http://www.Neip.info/downloads/l_bia1_resenha_santos.htm.

SANTOS, R.G. (2005). Review of: *Ordem, xamanismo e dádiva: o poder do Santo Daime, Labirinto, Revista Eletrônica do Centro de Estudos do Imaginário*, no. 7, available at: http://www.cei.unir.br/res51.html.

SANTOS, R.G. (2006a). Review of: *A reinvenção do uso da Ayahuasca nos centros urbanos, Terra Mística*, available at: http://www.terramistica.com.br/index.p hp?add=Artigos&file=article&sid=482&ch=1 and http://www.Neip.info/ downloads/l_bia2_res_rafa.htm.

SANTOS, R.G. (2006b). "Ensaio sobre a cura no contexto de um grupo da Barquinha," *Núcleo de Estudos Interdisciplinares sobre Psicoativos – Neip*, available at: http://www.neip.info/downloads/rafael/ensaio_cura.pdf.

SANTOS, R.G. (2006c). "Ayahuasca e redução do uso abusivo de psicoativos: eficácia terapêutica? *Núcleo de Estudos Interdisciplinares sobre Psicoativos – Neip*, available at: http://www.neip.info/downloads/rafael/texto_23_05. pdf. [Results of a case study of a youth who abandoned the abusive consumption of cocaine, alcohol, and tobacco after encountering the ritual use of ayahuasca in an independent group. Brasília (DF), 2006]

SANTOS, R.G. (2006d). "Efeitos da ingestão de Ayahuasca em estados psicométricos relacionados ao pânico, ansiedade e depressão em membros do culto Santo Daime," Mestrado em Psicologia – Processos Comportamentais. UnB. [Results of research conducted with long-term consumers of ayahuasca nine members of Santo Daime with at least ten years' experience). Brasília (DF), 2006], available at: http://www.neip.info/ downloads/rafael/ tese_rafa.pdf.

SANTOS, R.G. (2007). "Por uma abordagem multidisciplinar no estudo do consumo de psicoativos," *Núcleo de Estudos Interdisciplinares sobre Psicoativos – Neip*, available at: http://www.neip.info/downloads/rafael/abord_psic. pdf.

SANTOS, R.G.; MORAES, C.C; HOLANDA, A. (2006). "Ayahuasca e redução do uso abusivo de psicoativos: eficácia terapêutica?," *Psicologia: Teoria e Pesquisa*, no. 22, vol. 3, pp. 363-370, available at: http://www.scielo.br/scielo.php?script=sci_arttext&pid=S0102-

37722006000300014&lng=pt&nrm=iso&tlng=pt.

SHANON, B. (2003). "Os conteúdos das visões da Ayahuasca," *Mana*, no. 9, vol. 2, pp. 109-152. [Results of research with about 200 individual from indigenous groups and mestizo populations in Colombia, Ecuador, and Peru, and with members of the Barquinha, the União do Vegetal, Santo Daime, and of alternative groups in Brazil and Europe between 1994 and 2000]

SHANON, B. (2004). "A Ayahuasca e o estudo da mente," in: LABATE, B.C.; ARAÚJO, W.S. (eds.). *O uso ritual da Ayahuasca*. 2nd ed., Campinas, Mercado de Letras, pp. 681-709.

SILVA, L. O. da (2002). "A natureza do ritual do Santo Daime e a teoria de Victor Turner," *Último andar*, no. 7, pp. 103-114.

SILVA, L. O. da. Review of: A reinvenção do uso da Ayahuasca nos centros urbanos. *Rever: Revista de Estudos da Religião*, available at: http://www.pucsp.br/rever/resenha/labate01.htm and http://www.neip.info/downloads/resenha_leandro.htm.

SILVA, L. O. da (2004). "Marachimbé chegou foi para apurar: estudo sobre o castigo simbólico, ou peia, no culto do Santo Daime," Mestrado em Ciências da Religião, PUC-SP.

SILVA SÁ, D.B.G. (1996). "Ayahuasca, a consciência da expansão," *Discursos sediciosos – Crime, Direito e Sociedade*, no. 2, pp. 145-174, available at http://www.neip.info/downloads/domingos/domingos_bernardo.pdf.

SILVA SÁ, D.B.G. (2007). "A legitimidade jurídica do uso da Ayahauasca em rituais religiosos," VIII Fórum Internacional em Saúde: as drogas lícitas e ilícitas na Amazônia Legal: uso, abuso e recuperação, Rio Branco.

SILVEIRA, T.C.C. (2007). "Em busca da 'Santa Luz': um estudo sobre a experiência do êxtase místico-religioso na comunidade do Santo Daime," Mestrado em Ciências da Religião, UFJF, available at: http://alto-das-estrelas.blogspot.com/2007/07/dissertao-de-mestrado-em-andamento.html.

SOARES, L.E. (1994a). "Religioso por natureza: cultura alternativa e misticismo ecológico no Brasil," in: Soares, L.E. *O Rigor da Indisciplina. Ensaios de antropologia interpretativa.* Rio de Janeiro, Iser/Relume-Dumará, pp. 189-212.

SOARES, L.E. (1994b). "O Santo Daime no contexto da nova consciência religiosa," in: Soares, L.E. *O rigor da indisciplina. Ensaios de antropologia interpretativa.* Rio de Janeiro, Iser/Relume-Dumará, pp. 213-222, available at: http://www.neip.info/downloads/l_soares/daime_NCR.pdf.

SOARES, L.E. (1994c). "Misticismo e reflexão," in: Soares, L.E. *O rigor da indisciplina: ensaios de antropologia interpretativa.* Rio de Janeiro, Iser/Relume-Dumará, pp. 223-231.

TEIXEIRA, M.R.C. (2004). "Em roda dos meninos: um estudo da visão de mundo construída pelas crianças na cotidianidade da doutrina do Santo Daime na vila Céu Mapiá/AM," Mestrado em Educação, UFSC, available at: http://www.sbpcnet.org.br/livro/58ra/SENIOR/RESUMOS/resumo_213.html.

TEIXEIRA, M.R.C. (2005). "As crianças da floresta: na cotidianidade da doutrina do Santo Daime," Tomo 8.

TEIXEIRA, M.R.C. (2006). "Em roda dos meninos: um estudo da visão de mundo construída pelas crianças da floresta, na cotidianidade da doutrina do Santo Daime, na vila Céu Mapiá/AM," Anais da 58 Reunião Anual da SBPC, Florianópolis, available at: http://www.sbpcnet.org.br/livro/58ra/SENIOR/RESUMOS/resumo_213.html.

TROMBONI, M. (2003). Review of: *O uso ritual da Ayahuasca, Mana,* no. 9, vol. 2, pp. 211-215.

VIANA, T.C. (1997). *O consagrado defensor.* Belo Horizonte, Lítera Maciel Ltda.

Spanish

AGUIRRE, J.C. (coord.) (2000a). "Mesa redonda entre Luis Llorente y Tulio Cícero," Revista Monográfica *El idiota,* no. 1, pp. 236 y ss.

AGUIRRE, J.C. (coord.) (2000b). "Entrevista a Dacio Mingrone," *Revista Monográfica El idiota,* no. 1, pp. 255 y ss.

AGUIRRE, J.C. (2007). "Dacio Mingrone: un perfil," *Ulises,* no. 9.

ALVERGA, A.P. de (1994). "Ayahuasca: vida y enseñanzas del padrino Sebastian y el Santo Daime," Barcelona, Obelisco.

CAL OVEJERO, F. (1996). *Relatos del Santo Daime.* Madri, Ed. Amica. (Manuscript)

CAMARGO, I.A. (2003). "El uso religoso del té Ayahuasca y su relación con la psicosis: un estudio centralizado en la Unión del Vegetal y en el Santo Daime," Maestria en Drogodependencias, Universidad de Barcelona. [Results of observations made as a member of the União do Vegetal, by other members of the group in Brazil, and of some patients in the author's psychology clinic in Salvador (Bahia state) who began attending the UDV, as well as of members of the UDV who became clients of the author, 1996 to 2003], available at: http://alto-das-estrelas.blogspot.com/2005/05/tese-sobre-psicose-no-santo-daime-e-na.html.

CARNEIRO, H. (2004). *O uso ritual da Ayahuasca. Vision Chamanica.* Resenha, available at: http://www.visionchamanica.com/Publicaciones/index.htm.

CARVALHO, J.J. de (2001a). "El misticismo de los espíritus marginales," *Série Antropológica,* no. 294.

CARVALHO, J.J. de (2001b). "El misticismo de los espíritus marginales," *Marburg Journal of Religion,* no. 6, vol. 2, pp. 1-29.

FERICGLA, J.M. (2003). "La Ayahuasca en las nuevas religiones mistericas americanas," in: Fericgla, J.M. (ed.). *BI: Boletín de la Sociedad de Etnopsicología Aplicada. Monográfico Ayahuasca.* Barcelona, SdEA, pp. 18-33.

GONZÁLEZ, D. (2006). "La Ayahuasca," *Spannabis magazine,* no. 26, pp. 76-80.

GROISMAN, A. (1992). "Muerte y renacimiento: concepciones acerca de la espiritualidad de la muerte en la doctrina del Santo Daime," in: Cipoletti, M.S.; Langdon, E.J. (eds.). *La muerte y el "más allá" en las culturas indígenas latinoamericanas.* Quito, Ed. Abya-Yala, pp. 91-111.

HENMAN, A. (1986). "Uso del Ayahuasca en un contexto autoritario.

El caso de la União do Vegetal en Brasil." *America Indígena*, no. 46, vol. 1, pp. 219-234, available at: http://www.santodaime.it/Library/ANTROPOLOGY&SOCIOLOGY/henman86_spanish.htm.

IBARS, E.M. (2003). "De como llegué a conocer al Santo Daime y otros detalles," " in: Fericgla, J.M. (ed.). *BI: Boletín de la Sociedad de Etnopsicología Aplicada. Monográfico Ayahuasca*. Barcelona, SdEA, pp. 51-53.

LABATE, B.C. (2001). "Un panorama del uso ritual de la Ayahuasca en el Brasil contemporáneo," in: Mabit, J. (ed.). *Ética, mal y transgresión*. Lima, Takiwasi and Cisei, pp. 73-90, available at: http://www.neip.info/downloads/Tespan.pdf.

LAVAZZA, V.H. (2007). "Ideología y utópica en un culto milenarista. La experiencia del culto del Santo Daime en la Argentina," XIV Jornadas sobre Alternativas Religiosas en America Latina, Buenos Aires, Unsam, available at http://alto-das-estrelas.blogspot.com/2007/10/dia-26-de-setembro-de-2007-sandra.html.

LAVAZZA, V.H. (2007). "Comunidad y experiencia en un culto brasileño: los caminos del Santo Daime en Argentina," Mestrado em Antropologia Social, Universidad Nacional de San Martin, Buenos Aires, Argentina. (Project in development) http://alto-das-estrelas.blogspot.com/2007/09/aluno-de-mestrado-em-antropologia.html.

MacRAE, E. (1998). *Guiado por la luna*. Quito, Ed. Abya-Yala.

MacRAE, E. (2000). *El Santo Daime y la espiritualidad brasileña*. Ediciones Abya-Yala, Quito.

MAYORAL, F.R. (2005). "Santo Daime. La religión de la selva, en España," *CNR*, no. 104, pp. 92-96.

MILANEZ, W. (1999). *Oaska, el evangelio de la rosa*. Campinas, Sama Editora.

MONTEIRO, C.S. (1985). "La question de la realidad en la Amazônia: uma análisis a partir del estúdio de la doctrina Del Santo Daime," *Amazonia Peruana*, no. 6, vol. 11, pp. 87-100.

MONTEIRO, C.S. (1988). "Culto Del Santo Daime: chamanismo rural-urbano en Acre," in: Reichel, D. (ed.). *Rituales y fiestas de las Américas*. Bogotá, Ediciones Uniandes, pp. 286-300.

PERLONGHER, N. (n.d.). Luz de Cristal: fuerza y forma en la religión del Santo Daime. (Manuscript)

VILLAESCUSA, M. (2003). "Aspectos psicoterapéuticos de las ceremonias del Santo Daime en el Reino Unido," in: FERICGLA, J.M. (ed.). *BI: Boletín de la Sociedad de Etnopsicología Aplicada. Monográfico Ayahuasca*. Barcelona, Sd'EA, pp. 40-50.

WEISKOPF, J. (2002). *Yajé, el nuevo purgatorio*. Bogotá, Villegas Editores.

WEISKOPF, J. (2005). "Alabanza al Santo Daime. Carta a un ayahuasquero colombiano," *Núcleo de Estudos Interdisciplinares sobre Psicoativos – Neip*, available at: http://www.neip.info/ downloads/Texto%20Jimmy.pdf.

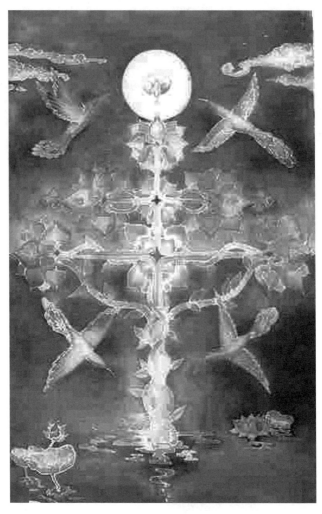

"Cruzeiro" by Daniel Mirante and Basienka Deerheart –
Batik, 90 cm by 160cm.
Credits: Daniel Mirante and Basienka Deerheart

About the Authors

Beatriz Caiuby Labate (Bia Labate) http://bialabate.net was born in 1971 in São Paulo, Brazil, where she lives. She earned a degree in social sciences from Unicamp in 1996, and in 2000 took the degree of Mestre in Antropologia Social from the same university. Her MA thesis earned the 2000 Prêmio de Melhor Tese de Mestrado em Ciências Sociais for Brazil's best social science thesis from ANPOCS (the National Association of Post-Graduate Studies in Social Sciences). She is a PhD student in Social Anthropology at Unicamp. She co-edited the books *O uso ritual da ayahuasca* (Mercado de Letras 2002, 2004 2nd ed.), *O uso ritual das plantas de poder* (Mercado de Letras, 2005), *Drogas e cultura: novas perspectivas* (EDUFBA, 2008) and *Ayahuasca y Salud* (Los Libros de La Liebre de Marzo, in press), and is also co-editor of a special edition of the journal *Fieldwork in Religion* titled *Light from the Forest: The ritual use of Ayahuasca in Brazil* (2006 published 2008) and author of *A reinvenção do uso da ayahuasca nos centros urbanos* (Mercado de Letras, 2004). She is a member of the Núcleo de Estudos Interdisciplinares sobre Psicoativos (NEIP) and is editor of its website, www.neip.info. Currently she works as an independent writer, consultant, lecturer and organizer of scientific conferences and cultural events related to the field of drugs, religion and shamanism. She is also a blogger and an anti-prohibitionist activist. blabate@bialabate.net

 **Isabel Santana de Rose** was born in 1980. She studied social science at Unicamp. As an undergraduate she began her studies of Santo Daime, writing a monograph entitled "A cura na doutrina do Santo Daime – Um estudo sobre o Céu da Mantiqueira" ("Healing in Santo Daime – A study of Céu da Mantiqueira"). She continued this research for her master's degree in social anthropology at the Universidade Federal de Santa Catarina, where she deepened her discussion of healing and therapeutic practices in Santo Daime; in 2005 she defended the thesis "Espiritualidade, terapia e cura – Um estudo sobre a expressão da experiência no Santo Daime" ("Spirituality, therapy, and healing – A study of the expression of experience in Santo Daime"). She is currently a PhD student at the same university, where she is researching the ways Guarani Indians from the Ynyn Moroti Wherá village (in Biguaçu, SC) are incorporating the use of ayahuasca in their ceremonies and in their discourse on culture and the Guarani tradition. She is also a researcher affiliated with the Núcleo de Estudos Interdisciplinares sobre Psicoativos (NEIP). belderose@yahoo.com.br

Rafael Guimarães dos Santos is a resident of Brasília, where he was born in 1980. In 2004 he earned a degree in biology from the Centro Universitário de Brasília – UniCEUB. His senior thesis was entitled "Ayahuasca: chá de uso religioso. Estudo microbiológico, observações comportamentais e estudo histomorfológico de cérebro em Murídeos (*Rattus norvegicus* da linhagem Wistar)" ("Ayahuasca: A tea used religiously. Microbiological studies, behavioral observations, and cerebral histomorphological studies in Murids (*Rattus norvegicus* of the Wistar strain)."  There, he analyzed the biological, behavioral, and cultural aspects of the use of ayahuasca. He has a master's degree in psychology from the Universidade de Brasília (UnB), where he researched the possible relationship between the ritualized use of ayahuasca and states of depression, anxiety, and panic. He is currently researching the interface between organic and religious-cultural aspects of hallucinogens in general. He is a PhD student in pharmacology at the Universidad Autónoma de Barcelona, under Dr. Jordi Riba. He is also a researcher affiliated with the Núcleo de Estudos Interdisciplinares sobre Psicoativos (NEIP). banisteria@gmail.com

"São João no Céu do Planalto," by Gervásio Santo Silva – screen ink on paper.
Photo by Manuel Poppe

About the Publisher

Founded in 1986, the Multidisciplinary Association for Psychedelic Studies (MAPS) is a membership-based, IRS-approved 501 (c) (3) non-profit research and educational organization. We assist scientists to design, fund, obtain approval for, conduct, and report on studies evaluating the risks and benefits of MDMA, psychedelic drugs, and marijuana. MAPS' mission is to sponsor scientific research designed to develop psychedelics and marijuana into FDA-approved prescription medicines and to educate the public honestly about the risks and benefits of these drugs.

For decades, the government was the biggest obstacle to research. Now that long-awaited research is finally being approved, the formidable challenge is funding it. At present, there is no funding available from governments, pharmaceutical companies, or major foundations. That means, for the time being, that the future of psychedelic and marijuana research rests in the hands of people like you.

Can you imagine a cultural reintegration of the use of psychedelics and the states of mind they engender? Please join MAPS in supporting the expansion of scientific knowledge in this promising area. Progress is only possible with the support of individuals who care enough to take individual and collective action.

Since 2000, MAPS has disbursed over four million dollars to worthy research and educational projects.

How MAPS Has Made a Difference

- Sponsored and obtained approval for the first LSD-assisted psychotherapy study in over 35 years. The study is taking place in Switzerland in subjects with anxiety associated with end-of-life issues.

- Sponsored the first US FDA-approved study evaluating MDMA's therapeutic applications, for subjects with chronic posttraumatic stress disorder (PTSD), as well as MDMA/PTSD pilot studies in Switzerland, Israel, Canada, and Spain.

- Waged a successful lawsuit against DEA in support of Professor Lyle Craker's proposed MAPS-sponsored medical marijuana production facility at the University of Massachusetts-Amherst; led campaigns to gain support from over 50 members of the US House of Representatives.

- Supported long-term follow-up studies of pioneering research with LSD and psilocybin from the 1950s and 1960s.

- Sponsored Dr. Evgeny Krupitsky's pioneering research into the use of ketamine-assisted psychotherapy in the treatment of alcoholism and heroin addiction.

- Assisted Dr. Charles Grob to obtain permission for the first human studies in the United States with MDMA after it was criminalized in 1985.

- Sponsored the first study to analyze the purity and potency of street samples of "Ecstasy" and medical marijuana.

- Funded the successful effort of Dr. Donald Abrams to obtain permission for the first human study into the therapeutic use of marijuana in 15 years, and to secure a million dollar grant from the National Institute on Drug Abuse.

- Obtained orphan-drug designation from the FDA for smoked marijuana in the treatment of AIDS Wasting Syndrome.

- Funded the synthesis of psilocybin for the first FDA-approved study in a patient population in twenty-five years.

- Sponsored "Psychedelic Harm Reduction" programs and services at events, concerts, schools, and churches.

Benefits of a MAPS Membership

MAPS members receive the tri-annual *MAPS Bulletin*. In addition to reporting on the latest research in both the United States and abroad, the *Bulletin* includes feature articles, personal accounts, book reviews, and reports on conferences and allied organizations. MAPS members are invited to participate in a vital on-line mailing list and to visit our website, which includes all articles published by MAPS since 1988.

Unless otherwise indicated, your donation will be considered an unrestricted gift to be used to fund high-priority projects. If you wish, however, you may direct contributions to a specific study. Your tax-deductible donations may be made by credit card or check made out to MAPS. Gifts of stock are welcome, as are trust and estate planning options.

The MAPS list is strictly confidential and not available for purchase. The *MAPS Bulletin* is mailed in a plain envelope.

MAPS Subscription Rates

Student/Low income Membership — $20
Basic Membership — $35
Integral Membership — $50 *(Includes a complimentary book)*
Supporting Membership — $100 *(Includes a complimentary book)*
Patron Membership — $250+ *(Includes two complimentary books)*
International members add $15 for postage

Multidisciplinary Association for Psychedelic Studies
309 Cedar Street #2323, Santa Cruz CA 95060
voice: (831)336-4325
fax: (831)336-3665
e-mail: askmaps@maps.org
Please visit our website: www.maps.org

Ordering Information

Ayahuasca Religions: A Comprehensive Bibliography and Critical Essays
(ISBN: 0-9798622-0-5) ...$11.95/copy

Shipping and Handling Charges

Domestic book rate (allow 3 weeks): $3.50
Domestic priority mail (allow 7 days): $5.00
($1.00 each additional copy)

Overseas airmail rates (allow 10 days):
Canada/Mexico ($9.00)
Other Countries ($12.00) — Pacific Rim ($15.00)
($1.00 each additional copy)

Methods of Payment

Check or money order in U.S. Dollars.
Mastercard, Visa, American Express
Wholesale Orders Welcome: Discount: 50%

Other ways to order:
Via secure credit card transaction at www.maps.org
Through your favorite local bookstore

Send Orders to:

MAPS
309 Cedar Street #2323, Santa Cruz CA 95060
voice: (831)336-4325 • fax: (831)336-3665
e-mail: orders@maps.org • www.maps.org

Suggested citation for this book:
Labate, B. C., Rose, I. S., & Santos, R. G. (2009).
Ayahuasca Religions: A Comprehensive Bibliography and Critical Essays.
Santa Cruz, CA: MAPS.

This book had the support of:
bialabate.net and NEIP.

http://bialabate.net

http://www.neip.info

PUBLISHED BY

http://www.maps.org